Revision Exercises

Revision Exercises

Based on Textbook of Biochemistry (7th Edition)

Essay Questions, Short Notes, Viva Voce and Multiple Choice Questions

DM Vasudevan MBBS MD FAMS FRCPath
Distinguished Professor
Department of Biochemistry
College of Medicine, Amrita Institute of Medical Sciences
Kochi, Kerala, India
Formerly
Principal, College of Medicine
Amrita Institute of Medical Sciences, Kerala, India
Dean, Sikkim Manipal Institute of Medical Sciences
Gangtok, Sikkim, India

Sreekumari S MBBS MD
Professor and Head
Department of Biochemistry
Sree Gokulam Medical College and Research Foundation
Thiruvananthapuram, Kerala, India

Kannan Vaidyanathan MBBS MD
Professor and Head
Department of Biochemistry
Pushpagiri Institute of Medical Sciences and Research Center
Thiruvalla, Kerala, India

JAYPEE BROTHERS MEDICAL PUBLISHERS (P) LTD
New Delhi • London • Philadelphia • Panama

Jaypee Brothers Medical Publishers (P) Ltd

Headquarters
Jaypee Brothers Medical Publishers (P) Ltd
4838/24, Ansari Road, Daryaganj
New Delhi 110 002, India
Phone: +91-11-43574357
Fax: +91-11-43574314
Email: jaypee@jaypeebrothers.com

Overseas Offices

J.P. Medical Ltd
83, Victoria Street, London
SW1H 0HW (UK)
Phone: +44-2031708910
Fax: +02-03-0086180
Email: info@jpmedpub.com

Jaypee-Highlights Medical Publishers Inc
City of Knowledge, Bld. 237, Clayton
Panama City, Panama
Phone: +507-301-0496
Fax: +507-301-0499
Email: cservice@jphmedical.com

Jaypee Medical Inc
The Bourse
111 South Independence Mall East
Suite 835, Philadelphia, PA 19106, USA
Phone: + 267-519-9789
Email: joe.rusko@jaypeebrothers.com

Jaypee Brothers Medical Publishers (P) Ltd
17/1-B Babar Road, Block-B, Shaymali
Mohammadpur, Dhaka-1207
Bangladesh
Mobile: +08801912003485
Email: jaypeedhaka@gmail.com

Jaypee Brothers Medical Publishers (P) Ltd
Shorakhute, Kathmandu
Nepal
Phone: +00977-9841528578
Email: jaypee.nepal@gmail.com

Website: www.jaypeebrothers.com
Website: www.jaypeedigital.com

Inquiries for bulk sales may be solicited at: jaypee@jaypeebrothers.com

Revision Exercises

First Edition: 1995

Seventh Edition: **2013**

ISBN 978-93-5090-637-8

Printed at: Ajanta Offset & Packagings Ltd., New Delhi

Preface

The booklet *Revision Exercises* containing various types of questions, was provided along with the previous editions of the textbook. It was found to be very popular among students, as it gave a glance of all topics covered, and useful for last-minute preparation for examinations.

Considering the feedbacks from students, this *Revision Exercises* booklet has been substantially updated. There are about 170 essay questions, 400 short note questions, 1000 multiple choice questions (MCQs) and 1,400 viva voce type questions in this booklet.

Question papers of various universities now contain MCQs; so it is necessary that students should familiarize with such questions. Therefore, a separate section is provided in this booklet. A lot of questions have been added to the previous edition, so that the present booklet contains about 1000 MCQs.

This Revision Exercises will help the students to identify the important points. Further, for the last-minute preparation before the examination, this companion booklet will be highly beneficial to the students. We are sure this question bank will be useful to the students and faculty members. Suggestions for improvement are most welcome.

DM Vasudevan
Sreekumari S
Kannan Vaidyanathan

Contents

Part 1: Essay Questions

CHAPTER 2: CELL

2-1. Enumerate the cellular organelle and list the functions of different organelles.
2-2. Give a labeled diagram of mitochondria and enumerate the functions.
2-3. What are the differences between eukaryotic and prokaryotic cells?
2-4. Briefly outline the structure of biomembrane with the help of a diagram.
2-5. Enumerate different transport mechanisms with suitable examples.
2-6. Define active transport. Explain the different types of active transport with examples.
2-7. Enumerate the salient features of active transport. Give two examples where drugs inhibit active transport.
2-8. Describe the different types of ion channels.

CHAPTER 3: AMINO ACIDS, STRUCTURE

3-1. Classify amino acids, giving suitable examples.
3-2. How will you classify amino acids based on their nutritional importance?
3-3. Define iso-electric pH and give the importance.
3-4. Enumerate different reactions of amino acids, giving one example for each.
3-5. Give an account of the transamination reactions. Give two suitable examples. What is the metabolic importance of transamination? What is the clinical application of transaminase estimation?

CHAPTER 4: PROTEINS, STRUCTURE

4-1. How does the charge on a protein molecule vary with the change in pH of the medium?
4-2. What is end group analysis? What are the reagents used for this purpose?
4-3. Describe the primary, secondary and tertiary structures of proteins. What are the forces, which stabilize them?
4-4. What is the primary structure of a protein? Explain with the help of the structure of Insulin.
4-5. Name two common types of secondary structures. Mention how they are preserved.

4-6. Name two proteins with quaternary structure. Indicate their subunit make up. How are they maintained?

4-7. Explain the structural organization of hemoglobin molecule. How does the alteration in amino acid sequence affect the properties of hemoglobin?

4-8. What are the different techniques used for precipitation of proteins?

4-9. Classify proteins with suitable examples.

4-10. Enumerate the different techniques available for the estimation of proteins. Explain the principle of any one of them.

CHAPTER 5: ENZYMES, GENERAL CONCEPTS

5-1. Enumerate the major classes of enzymes, giving one example for each.

5-2. What are the salient features of the active site of the enzyme?

5-3. Explain in detail the factors affecting the velocity of an enzyme reaction.

5-4. What are the different types of enzyme inhibition? Explain with suitable examples.

5-5. What is competitive inhibition? Explain with two examples of its therapeutic significance.

5-6. What are the differences between competitive and non-competitive inhibition? Give two examples for competitive inhibition.

5-7. What is meant by Km value, and what is its significance?

5-8. Explain the mechanism of allosteric regulation using phosphofructokinase as an example.

5-9. With suitable examples, explain the specificity of enzymes.

5-10. What is the importance of covalent modification of enzymes?

5-11. What are iso-enzymes? Give examples. What are their clinical significance?

CHAPTER 7: LIPIDS, CHEMISTRY

7-1. Classify lipids, giving examples.

7-2. What are phospholipids? Name their functions.

CHAPTER 9: GLUCOSE, MAJOR PATHWAYS

9-1. Describe digestion and absorption of carbohydrates.

9-2. What is the major catabolic pathway of glucose under anaerobic conditions? Mention the steps in the pathway and indicate the key enzymes.

9-3. Describe the process of glycolysis. Explain how many molecules of ATP are formed in anaerobic and aerobic conditions.

9-4. Assuming that oxygen is available in plenty, what are the steps in which energy is trapped in glycolytic pathway? Give the substrate, product, enzyme and coenzyme of each of these energy-trapping steps.

9-5. How is glucose metabolized by the RBCs for producing energy? Indicate the energy yield from 1 molecule of glucose in the RBCs.

9-6. In anaerobic glycolysis, lactic acid is generated. What is the reason for reduction of pyruvate to lactate?

9-7. What are the irreversible steps in glycolysis? How are these blocks circumvented?

9-8. Trace the pathway of gluconeogenesis starting from alanine. Mention the key enzymes.

9-9. "Gluconeogenesis is not a simple reversal of glycolysis"; justify your answer by giving the reactions.

CHAPTER 10: OTHER PATHWAYS OF GLUCOSE

10-1. How is glycogen broken down in the body? Explain the hormonal regulation of the pathway.

10-2. Outline the sequence of reactions involved in the breakdown of glycogen in the skeletal muscle. Explain the purpose served by this process.

10-3. Write the reactions of the oxidative phase of the hexose monophosphate shunt pathway. Which tissues have this pathway?

10-4. What is the significance of HMP shunt pathway?

CHAPTER 11: OTHER CARBOHYDRATES

11-1. Describe the process by which galactose is converted into glucose. Indicate the metabolic errors associated with this pathway.

CHAPTER 12: METABOLISM OF FATTY ACIDS

12-1. Classify lipids, giving examples.

12-2. Name the dietary lipids. Explain digestion and absorption of fats.

12-3. Explain the role of bile salts in the digestion and absorption of dietary lipids. Mention the changes observed in obstructive jaundice.

12-4. Explain the steps of beta oxidation of palmitic acid, giving energetics.

12-5. Describe the beta oxidation of fatty acids. What are the co-enzymes required for beta oxidation, and in which steps?

12-6. Give the sources and fate of Acetyl-CoA.

12-7. Describe the reactions of de novo synthesis of fatty acid and how is it regulated?

12-8. Describe the de novo synthesis of fatty acids. What is the co-enzyme required, and how is it generated?

12-9. How are the fatty acids in adipose tissue mobilized and transported to other tissues? Explain the effect of hormones in this process.

12-10. How is propionyl CoA formed? How is it further metabolized?

12-11. What is fatty liver? Explain the causes of fatty liver. Indicate how lipotropic factors can prevent fatty liver.

12-12. What are ketone bodies? Explain the reactions leading to the formation of them. How are they utilized in the body?

12-13. Name the ketone bodies. Give two conditions characterized by excessive production of ketone bodies. Explain the metabolic derangements and consequences of ketosis.

CHAPTER 13: CHOLESTEROL AND LIPOPROTEINS

13-1 Enumerate the major steps of synthesis of cholesterol. Name the rate-limiting step of cholesterol synthesis. Describe the role of cholesterol in atherosclerosis.

13-2. Classify lipoproteins. Explain their biological significance.

13-3. What is the normal cholesterol level in plasma? Explain how the cholesterol is transported from liver to peripheral tissues and back?

13-4. How are dietary triglycerides absorbed and transported in plasma? Explain briefly the transport of dietary TAG from intestine to liver.

13-5. How are endogenously produced triglycerides transported in plasma? Explain the transport of them from liver. Indicate the abnormalities associated with this process.

CHAPTER 14: PUFA, PROSTAGLANDINS

14-1. How are prostaglandins synthesized? Indicate their importance as local hormones. Mention the effects of drugs acting on this pathway.

CHAPTER 15: UREA CYCLE

15-1. Explain the term transamination. Give one suitable example. What is the metabolic importance of transamination? What is the clinical application of transaminase estimation?

15-2. Describe the reactions of the urea cycle. Discuss the interrelation of urea cycle and citric acid cycle.

15-3. Give details of the steps by which ammonia is detoxified in the brain and in liver.

15-4. What is uremia? Describe the reactions of urea cycle. Discuss the diagnostic significance of blood urea.

CHAPTER 16: SIMPLE AMINO ACIDS

16-1. What are the important substances derived from glycine? Describe the steps by which creatine is synthesized.

16-2. Write the reactions by which glycine is synthesized and catabolized. Name six important compounds derived from glycine and indicate their functions.

16-3. Describe the metabolism of methionine. Explain the term transmethylation with suitable examples.

16-4. Describe the fate and functions of methionine and cysteine.
16-5. Describe the pathway of methionine metabolism.
16-6. What is the biochemical basis of homocystinuria? What test you will do to diagnose homocystinuria?

CHAPTER 18: AROMATIC AMINO ACIDS

18-1. Describe the steps of catabolism of phenyl alanine and tyrosine. Indicate the inborn errors of metabolism associated with this pathway.
18-2. Describe the steps by which catecholamines are synthesized. What is the final excretory product of catecholamines?

CHAPTER 19: CITRIC ACID CYCLE

19-1. Discuss the formation of acetyl CoA from pyruvate. How is acetyl CoA is further metabolized in the citric acid cycle?
19-2. Give an account of the citric acid cycle and explain why it is called the common terminal metabolic pathway.
19-3. Write the steps of citric acid cycle. What is the biological significance of this cycle?
19-4. Describe the reactions of the citric acid cycle. How many ATP molecules are generated per molecule of acetyl CoA entering the cycle?
19-5. Explain the amphibolic nature of citric acid cycle.

CHAPTER 20: BIOLOGICAL OXIDATION

20-1. Write the members of the electron transport chain, in the order of redox potentials, and show the steps where ATP is synthesized.
20-2. Define oxidative phosphorylation. Explain the chemiosmotic theory.
20-3. Discuss metabolic alterations in starvation.

CHAPTER 21: HEME

21-1. Give an account of heme synthesis.
21-2. What is porphyria? Classify different types of porphyrias. Give an account of acute intermittent porphyria.
21-3. Describe the catabolism of heme in the body.

21-4. How bile pigments are formed? Give the significance of their presence in blood and urine. How are they detected in blood and urine?

21-5. Classify jaundice. How do you investigate a case of jaundice?

21-6. Discuss the biochemical alterations seen in blood and urine in different types of jaundice.

21-7. How is bilirubin formed in the body? Describe how it is excreted? Describe the biochemical changes in hepatocellular jaundice and obstructive jaundice.

CHAPTER 22: HEMOGLOBIN

22-1. What is the difference between hemoglobinopathies and thalassemias? Describe any one of the hemoglobinopathies in detail.

22-2. Give biochemical explanation for the finding that geographical distribution of glucose-6-phosphate dehydrogenase deficiency correlates well with malarial incidence.

CHAPTER 23: CLINICAL ENZYMOLOGY, BIOMARKERS

23-1. Write briefly about the enzymes that show variations in serum level in myocardial infarction, and their clinical significance.

23-2. Name the enzymes that are used in the diagnosis of myocardial infarction. Explain the pattern of rise following an acute infarction.

23-3. Indicate the importance of enzyme studies in (a) Myocardial infarction; (b) Liver diseases; (c) Bone diseases; (d) Malignancy.

23-4. Describe the importance of serum enzyme estimations in clinical diagnosis.

23-5. Indicate the use of enzyme studies in (a) Hepatobiliary diseases; (b) Bone diseases; (c) Muscle diseases; and (d) Pancreatic diseases

CHAPTER 24: REGULATION OF BLOOD GLUCOSE

24-1. What is the normal fasting blood glucose level? How is it regulated?

24-2. What are the hormones influencing blood sugar level and how are these hormones acting?

24-3. Give the normal blood glucose levels in fasting conditions and after a carbohydrate meal. What are the mechanisms by which normal range is maintained?

24-4. Discuss the changes in metabolism during diabetes mellitus.

24-5. What are the enzymes influenced by insulin? What are the derangements seen in diabetes mellitus?

24-6. What are the indications of glucose tolerance test? What precautions are to be taken before doing a GTT? What are the abnormal curves obtained? What is impaired glucose tolerance?

24-7. Name the reducing sugars that may appear in urine and give the differential diagnosis of these clinical conditions.

CHAPTER 25: CARDIOVASCULAR DISEASES

25-1. What is the normal cholesterol level in plasma? What is its clinical significance? What are the dietary precautions to reduce hypercholesterolemia?

CHAPTER 26: LIVER FUNCTION TESTS

26-1. Enumerate liver function tests and describe in detail any two of them with clinical significance.

26-2. Classify jaundice. Give an account of the biochemical tests which will help in differentiating the types of jaundice.

26-3. Name the bile pigments. Give the significance of their presence in blood and urine. How are they detected in blood and urine?

26-4. Classify jaundice. How do you investigate a case of jaundice?

26-5. Discuss the biochemical alterations seen in blood and urine in different types of jaundice.

CHAPTER 27: KIDNEY FUNCTION TESTS

27-1. Name the renal clearance tests. Give details of any one of them.

27-2. How creatinine clearance test is done? What is its diagnostic significance?

27-3. How urea clearance test is done? What is its clinical significance?

CHAPTER 28: PLASMA PROTEINS

28-1. Enumerate the major transport proteins of plasma. Explain the transport of free fatty acids, bilirubin, iron and calcium.

28-2. What are the important functions of albumin? Give the major causes and manifestations of hypoalbuminemia.

28-3. Indicate the electrophoresis abnormalities observed in the following conditions: (a) Cirrhosis liver; (b) Nephrotic syndrome and (c) Multiple myeloma.

CHAPTER 29: ACID-BASE BALANCE

29-1. What is the normal pH of blood? Explain the role of plasma buffers and renal mechanisms in the maintenance of acid-base balance of the body.

29-2. Name the important buffer systems in the body. Describe the role of kidney and lungs in the maintenance of acid-base balance.

29-3. What is titratable acidity of urine? What is the role of kidney in maintaining acid-base balance?

29-4. What is metabolic acidosis? Enumerate its causes. What are the compensatory mechanisms?

CHAPTER 30: ELECTROLYTE AND WATER BALANCE

30-1. Discuss the hormonal control of fluid and electrolyte homeostasis.

CHAPTER 35: SEPARATION AND QUANTITATION

35-1. Give the salient features of electrophoresis. What are the abnormalities that you could detect in serum electrophoresis?

35-2. Describe the principle and applications of ELISA.

CHAPTER 36: FAT SOLUBLE VITAMINS

36-1. Describe sources, biochemical functions, requirement and deficiency manifestations of vitamin A.

36-2. Describe the sources, biochemical functions, normal requirement and deficiency manifestations of Vitamin D.

CHAPTER 37: WATER SOLUBLE VITAMINS

37-1. Describe the source, biochemical functions, normal requirement and deficiency manifestations of thiamine.

37-2. Describe the sources, biochemical functions, normal requirement and deficiency manifestations of pyridoxal phosphate.

CHAPTER 38: WATER SOLUBLE VITAMINS CONTINUED

38-1. Describe sources, biochemical functions, requirement and deficiency manifestations of folic acid

38-2. Describe sources, biochemical functions, requirement and deficiency manifestations of vitamin B_{12}

38-3. Describe sources, biochemical functions, requirement and deficiency manifestations of Vitamin C.

CHAPTER 39: MINERAL METABOLISM

39-1. What is the normal blood level of calcium? What are the mechanisms by which calcium homeostasis is maintained?

39-2. Describe the sources, daily requirement, absorption, biochemical functions and deficiency manifestations of iron.

39-3. Name any four trace elements and mention the biological functions of each of them.

CHAPTER 40: NUTRITION

40-1. Define BMR. What are the factors that affect BMR?

40-2. What is a balance diet? How do you prepare a diet for a normal young adult male of sedentary habits?

40-3. What is the nutritional importance of dietary proteins? Explain how the dietary deficiency of proteins will affect growing children.

40-4. An adult male is consuming 350 g carbohydrate, 70 g protein and 60 g lipid per day. Calculate the total caloric content of his diet. Calculate the percentage of calories obtained from carbohydrates, lipids and proteins. Are the percentages of calories that have obtained from these macronutrients conforming to the requirements of a balanced diet?

CHAPTER 41: DETOXIFICATION

41-1. What are xenobiotics? Describe the role of glutathione in detoxification.

42-2. What is meant by detoxification? Give an account of various detoxification processes.

CHAPTER 43: NUCLEOTIDES

43-1. Give the sources of carbon and nitrogen atoms of purine and pyrimidine rings. How is the de novo synthesis regulated? Indicate the clinical uses of inhibitors of purine nucleotide synthesis.

43-2. Describe the inborn errors of metabolism associated with degradation pathways of purines and pyrimidines.

43-3. How are purine nucleotides degraded? Add a note on abnormalities due to excessive purine catabolism.

CHAPTER 44: DNA STRUCTURE, REPLICATION

44-1. Describe the structure of DNA. What are the differences between DNA and RNA?

44-2. What are the salient features of Watson-Crick model of DNA?

44-3. Describe the process of DNA replication. Name two inhibitors of replication.

CHAPTER 45: TRANSCRIPTION

45-1. Describe different types of RNA.

45-2. Give a detailed account of the transcription process. How is it regulated? Name inhibitors of transcription.

CHAPTER 46: TRANSLATION

46-1. What is a codon? Describe the salient features of genetic code.

46-2. Describe the phases of activation, initiation, elongation and termination of biosynthesis of protein.

46-3. Describe the steps of protein synthesis.

CHAPTER 47: MUTATIONS AND GENE EXPRESSION

47-1. Explain with suitable examples, how mutations result in abnormal proteins.

47-2. What is mutation? What are mutagens? Describe point mutation and frameshift mutation.

CHAPTER 48: RECOMBINANT DNA TECHNOLOGY

48-1. Describe recombinant DNA technology. What are the important applications of the technique?

48-2. How the DNA molecule cloned? Write two applications of DNA cloning.

CHAPTER 50: HORMONES, MECHANISM

50-1. What is cyclic AMP? What is its metabolic importance?

CHAPTER 53: THYROID HORMONES

53-1. Describe the synthesis and secretion of thyroxine.

53-2. Enumerate the thyroid function tests. Describe any one of them in detail.

CHAPTER 59: RADIOISOTOPES

59-1. Give two examples of isotopes and its applications in medicine.

59-2. What are the medical uses of isotopes?

Part 2: Questions for Short Notes

CHAPTER 2: CELL

2-1. Structure of cell membrane.
2-2. Fluid mosaic model.
2-3. Active transport.
2-4. Facilitated transport.
2-5. Symport system.
2-6. Name two diseases related to the lysosomal function.

CHAPTER 3: AMINO ACIDS, STRUCTURE

3-1. Decarboxylation of amino acids.
3-2. Name the branched chain amino acids.
3-3. What are essential amino acids?

CHAPTER 4: PROTEINS, STRUCTURE

4-1. Primary structure of proteins.
4-2. Alpha helix of proteins.
4-3. Tertiary structure of proteins.
4-4. Quaternary structure of proteins.
4-5. Denaturation of proteins.
4-6. Conjugated proteins.
4-7. Iso-electric pH.
4-8. Iso-electric precipitation.
4-9. Precipitation reactions of proteins.
4-10. Heat coagulation.
4-11. Decarboxylation of amino acids.

CHAPTER 5: ENZYMES, GENERAL CONCEPTS

5-1. Oxidoreductases.
5-2. Co-enzymes.
5-3. Metallo-enzymes.
5-4. Active site of enzyme.
5-5. Koshland's induced fit theory
5-6. Pro-enzymes.
5-7. Zymogens.
5-8. Effect of pH on enzyme activity.
5-9. Optimum temperature.
5-10. Michaelis Constant (Km).

CHAPTER 6: CARBOHYDRATES, CHEMISTRY

6-1. Classification of monosaccharides.
6-2. Reducing disaccharides.

6-3. Non-reducing disaccharide.
6-4. Why sucrose is a non-reducing sugar?
6-5. Transport mechanisms of glucose.
6-6. Mutarotation.
6-7. Anomers.
6-8. Epimerism.

CHAPTER 7: LIPIDS, CHEMISTRY

7-1. Rancidity.
7-2. Structure and function of lecithin.
7-3. Gangliosides.
7-4. Cerebrosides.

CHAPTER 9: GLUCOSE, MAJOR PATHWAYS

9-1. Regulation of glycolysis.
9-2. Key enzymes of glycolysis.
9-3. Phosphofructokinase.
9-4. Substrate level phosphorylation.
9-5. 2,3-bisphosphoglycerate (2,3-BPG).
9-6. Cori's cycle.
9-7. Substrates for gluconeogenesis.
9-8. Regulation of gluconeogenesis.
9-9. Key enzymes of gluconeogenesis.
9-10. Malate shuttle.

CHAPTER 10: OTHER PATHWAYS OF GLUCOSE

10-1. Action of glucagon on glycogenolysis.
10-2. Cyclic AMP.
10-3. Von-Gierke's disease.
10-4. Glycogen storage diseases.
10-5. Significance of HMP shunt pathway.
10-6. Transketolase.
10-7. Key enzyme of HMP shunt pathway.
10-8. Drug-induced hemolytic anemia.
10-9. UDP-glucuronic acid.

CHAPTER 11: MINOR PATHWAYS OF CARBOHYDRATES

11-1. Enzyme defect in galactosemia.
11-2. Enzyme defect in congenital cataract.
11-3. Lactose intolerance.
11-4. Hereditary fructose intolerance.

11-5. Essential pentosuria.
11-6. Metabolism of alcohol.

CHAPTER 12: METABOLISM OF FATTY ACIDS

12-1. Rate limiting enzyme of fatty acid biosynthesis.
12-2. Carnitine.
12-3. Role of citrate in fatty acid synthesis.
12-4. ATP citrate lyase.
12-5. Oxidation of odd chain fatty acids.
12-6. Metabolism of propionyl CoA.
12-7. Refsum's disease.
12-8. Alpha oxidation.
12-9. Dicarboxylic aciduria.
12-10. Effect of insulin on lipolysis.
12-11. Hormone sensitive lipase.
12-12. Fatty liver.
12-13. Lipotropic factors.
12-14. Ketosis.
12-15. Ketogenesis.
12-16. Give the enzyme defect in the following conditions: (a) Methyl malonyl aciduria; (b) Refsum's disease

CHAPTER 13: CHOLESTEROL AND LIPOPROTEINS

13-1. Biologically important compounds derived from cholesterol.
13-2. HMG CoA reductase.
13-3. Regulation of cholesterol synthesis.
13-4. Key enzymes of cholesterol biosynthesis.
13-5. Give the normal blood level of cholesterol.
13-6. Apo-B-100.
13-7. Chylomicrons.
13-8. Lipoprotein lipase.

CHAPTER 14: PUFA, PROSTAGLANDINS

14-1. Prostaglandins.
14-2. Cyclo-oxygenase.
14-3. Mechanism of action of aspirin.
14-4. Effect of prostaglandin on smooth muscles.
14-5. Lipid storage diseases.
14-6. Gaucher's disease.
14-7. Niemann-Pick disease.
14-8. Tay-Sachs disease.

CHAPTER 15: UREA CYCLE

15-1. Proteolytic enzymes of gastrointestinal tract.
15-2. Digestion of proteins.
15-3. Transamination.
15-4. Transdeamination.
15-5. Name one-carbon units.
15-6. Name one-carbon donor.
15-7. Name one-carbon acceptors.
15-8. Decarboxylation of amino acids.

CHAPTER 16: SIMPLE AMINO ACIDS

16-1. Name the important compounds formed from glycine.
16-2. Creatine synthesis.
16-3. Creatinuria.
16-4. Biological action of glutathione.
16-5. Gamma aminobutyric acid (GABA).
16-6. Homocystinuria.
16-7. Cystathionuria.
16-8. Cystinuria.
16-9. Give four examples of transmethylation reactions.
16-10. S-adenosyl methionine (active methionine).
16-11. Give the enzyme defect in the following conditions:(a) Methyl malonyl aciduria; (b) Cystathionuria; (c) Homocystinuria.

CHAPTER 17: BRANCHED CHAIN AMINO ACIDS

17-1. Maple syrup urine disease.

CHAPTER 18: AROMATIC AMINO ACIDS

18-1. Phenyl ketonuria.
18-2. Alkaptonuria.
18-3. Albinism.
18-4. Important compounds derived from tyrosine.
18-5. Vanillyl mandelic acid (VMA).
18-6. Synthesis of catecholamines.
18-7. Hydroxy indole acetic acid (HIAA).
18-8. Synthesis and catabolism of serotonin.
18-9. Carcinoid syndrome.
18-10. Hartnup disease.
18-11. Name the ketogenic amino acids.

18-12. Give the enzyme defect in the following conditions:(a) Phenyl ketonuria; (b) Alkaptonuria; (c) Albinism; (d) Maple syrup urine disease.
18-13. Figlu excretion test.
18-14. Histamine.

CHAPTER 19: CITRIC ACID CYCLE

19-1. Energy releasing steps of citric acid cycle.
19-2. Carbon dioxide releasing steps during oxidation of glucose.

CHAPTER 20: BIOLOGICAL OXIDATION

20-1. Energy rich compounds.
20-2. Inhibitors of ETC.
20-3. Oxidative phosphorylation.
20-4. Uncouplers of oxidative phosphorylation.
20-5. Chemiosmotic theory.
20-6. Cytochromes.
20-7. Cytochrome oxidase.
20-8. ATP synthase.

CHAPTER 21: HEME

21-1. ALA synthase.
21-2. Acute intermittent porphyria.
21-3. Rate limiting step of heme synthesis.
21-4. Regulation of heme synthesis.
21-5. Formation and fate of bilirubin.
21-6. Catabolism of heme
21-7. Shunt bilirubin.
21-8. Crigler-Najjar syndrome.
21-9. van den Bergh reaction.
21-10. Hemolytic jaundice.
21-11. Hepatocellular jaundice.
21-12. Obstructive jaundice.
21-13. Urobilinogen.
21-14. Give the enzyme defect in the following conditions: (a) Drug-induced hemolytic anemia; (b) Crigler-Najjar syndrome.

CHAPTER 22: HEMOGLOBIN

22-1. Chloride shift.
22-2. Isohydric transport of carbon dioxide.

22-3. Oxygen dissociation curve.
22-4. Abnormal hemoglobins.
22-5. Hemoglobinopathies.
22-6. Sickle cell anemia.
22-7. Hemoglobin S.
22-8. Hemoglobin F.
22-9. Thalassemia.

CHAPTER 23: CLINICAL ENZYMOLOGY, BIOMARKERS

23-1. Clinical significance of troponins.
23-2. Clinical significance of CK.
23-3. Enzyme profile in myocardial infarction.
23-4. Enzyme profile in liver diseases.
23-5. Clinical significance of transaminases.
23-6. Clinical significance of ALP.
23-7. Clinical significance of ACP.
23-8. Glucose-6-phosphate dehydrogenase defect
23-9. Clinical significance of serum amylase.
23-10. Therapeutic uses of enzymes.
23-11. Immobilized enzymes.

CHAPTER 24: REGULATION OF BLOOD GLUCOSE

24-1. Key enzymes influenced by insulin.
24-2. Give the normal blood level of glucose
24-3. Renal glycosuria.
24-4. Renal threshold for glucose.
24-5. Benedict's test.
24-6. Glucose tolerance test (GTT).
24-7. Reducing sugars in urine.
24-8. Glycated hemoglobin.
24-9. HbA1c.

CHAPTER 25: CARDIOVASCULAR DISEASES

25.1. HDL-cholesterol.
25-2. LDL-cholesterol.
25-3. Prevention of atherosclerosis.
25-4. Polyunsaturated fatty acids.

CHAPTER 26: LIVER FUNCTION TESTS

26-1. Enzymes used as liver function tests.
26-2. Albumin-globulin ratio.

26-3. van den Bergh test.
26-4. Clinical significance of serum bilirubin level.
26-5. Fractional test meal.
26-6. Hyperchlorhydria.

CHAPTER 27: KIDNEY FUNCTION TESTS

27-1. Glomerular filtration rate.
27-2. Urine analysis
27-3. Proteinuria
27-4. Inulin clearance test.
27-5. Creatinine clearance test.
27-6. Creatinine coefficient.
27-7. Urea clearance test.
27-8. Give the normal blood level of the following: (a) Glucose; (b) Albumin; (c) Globulins; (d) Total protein; (e) Creatinine; (f) Urea; (g) Chloride; (h) Bicarbonate; (i) Sodium; (j) Potassium.

CHAPTER 28: PLASMA PROTEINS

28-1. Albumin-globulin ratio
28-2. Enumerate transport proteins of blood.
28-3. Ceruloplasmin.
28-4. Haptoglobin.
28-5. Transferrin.
28-6. Alpha-1 antitrypsin deficiency.

CHAPTER 29: ACID-BASE BALANCE

29-1. Bicarbonate buffer system of blood.
29-2. Alkali reserve.
29-3. Renal acidification of urine.
29-4. Renal production and excretion of ammonia.
29-5. Urinary ammonia.
29-6. Role of kidney in the regulation of pH.
29-7. Anion gap.
29-8. Metabolic acidosis.
29-9. Respiratory acidosis.
29-10. Metabolic alkalosis.
29-11. Respiratory alkalosis.
29-12. Give the normal blood level of the following: (a) Chloride; (b) Bicarbonate; (c) Sodium; (d) Potassium; (e) pH.

CHAPTER 30: ELECTROLYTE AND WATER BALANCE

30-1. Renin-angiotensn system
30-2. Isotonic contraction of ECF
30-3. Hypotonic contraction of ECF
30-4. Hypertonic contraction of ECF
30-5. Isotonic expansion of ECF

CHAPTER 31: BODY FLUIDS

31-1. Proteins in milk
31-2. Minerals in milk
31-3. Biochemical analysis of CSF

CHAPTER 33: FREE RADICALS AND ANTIOXIDANTS

33-1. Detoxification of hydrogen peroxide.
33-2. Superoxide dismutase.
33-3. Antioxidants.
33-4. Free radical scavenger mechanisms.
33-5. Reactive oxygen species (free radicals).

CHAPTER 34: CLINICAL LABORATORY

34-1. Prenatal diagnosis
34-2. Newborn screening
34-3. Tandom mass spectrometry
34-4. Preanalytical variables
34-5. Quality control

CHAPTER 35: SEPARATION AND QUANTITATION

35-1. Electrophoresis.
35-2. Paper chromatography.
35-3. Thin layer chromatography.
35-4. Affinity chromatography.
35-5. Radioimmunoassay.
35-6. ELISA (enzyme-linked immunosorbent assay).
35-7. Spectrophotometer
35-8. Dry chemistry systems

CHAPTER 36: FAT SOLUBLE VITAMINS

36-1. Sources and daily requirement of Vitamin A.
36-2. Functions of vitamin A.
36-3. Visual cycle.
36-4. Hypervitaminosis A.
36-5. Pro-vitamins.

36-6. Anti-vitamins.
36-7. Functions of vitamin D.
36-8. Activation of vitamin D.
36-9. Vitamin D deficiency.
36-10. Renal rickets.
36-11. Tocopherol.
36-12. Biological role of vitamin K.
36-13. Anti-coagulants.
36-14. Gamma carboxylation reaction.
36-15. What is the normal daily requirement of: (a) Vitamin A; (b) Vitamin D

CHAPTER 37: WATER SOLUBLE VITAMINS

37-1. Functions of thiamine pyrophosphate.
37-2. Deficiency of thiamine.
37-3. Beriberi.
37-4. Metabolic role of riboflavin.
37-5. Co-enzyme function of niacin.
37-6. Pellagra.
37-7. Functions of pyridoxal phosphate.
37-8. Biotin requiring reactions.
37-9. Co-enzyme functions of biotin.
37-10. Avidin.

CHAPTER 38: WATER SOLUBLE VITAMINS CONTINUED

38-1. Deficiency of folic acid.
38-2. Folate trap.
38-3. Figlu excretion test.
38-4. Folate antagonists.
38-5. Deficiency of vitamin B_{12}.
38-6. Biological role of vitamin B_{12}.
38-7. Absorption of vitamin B_{12}.
38-8. Intrinsic factor of Castle.
38-9. Methyl malonyl aciduria.
38-10. What is the normal daily requirement of: (a) Folic acid; (b) Vitamin B_{12}; (c) Vitamin C.

CHAPTER 39: MINERAL METABOLISM

39-1. Factors influencing calcium absorption.
39-2. Homeostasis of blood calcium level.

39-3. Hemosiderosis.
39-4. Absorption of iron.
39-5. Transferrin.
39-6. Hepcidin.
39-7. Causes and deficiency manifestations of iron.
39-8. Copper metabolism.
39-9. Ceruloplasmin.
39-10. Wilson's hepatolenticular degeneration.
39-11. Hyperkalemia.
39-12. Hypokalemia.
39-13. Functions of sodium.
39-14. Metabolic role of zinc.
39-15. Metabolic role of selenium.
39-16. Fluorosis.
39-17. What is the normal daily requirement of (a) Calcium; (b) Phosphorus; (c) Iron.

CHAPTER 40: NUTRITION

40-1. Basal metabolic rate.
40-2. Specific dynamic action.
40-3. Biological value of proteins.
40-4. Protein efficiency ratio.
40-5. Nitrogen balance.
40-6. Dietary fiber.
40-7. Essential amino acids.
40-8. Essential fatty acids.
40-9. Protein sparing effect of carbohydrates.
40-10. Protein calorie malnutrition.
40-11. Kwashiorkor.

CHAPTER 41: DETOXIFICATION

41-1. Metabolic role of glucuronic acid.
41-2. Detoxification by conjugation.
41-3. Detoxification by oxidation.
41-4. Detoxification by hydrolysis.
41-5. Detoxification of aspirin.
41-6. Detoxification of benzoic acid.
41-7. Detoxification of bilirubin.
41-8. Hippuric acid.
41-9. How is each of the following compounds detoxified: (a) Aspirin; (b) Ethyl alcohol; (c) Bilirubin; (d) Barbiturates.

CHAPTER 42: HEAVY METAL POISONS

42-1. Lead poisoning
42-2. Mercury poisoning
42-3. Aluminium toxicity.

CHAPTER 43: NUCLEOTIDES

43-1. Sources of carbon and nitrogen atoms in purine ring.
43-2. PRPP (phosphoribosyl pyrophosphate).
43-3. Regulation of purine synthesis.
43-4. Salvage pathway of purine synthesis.
43-5. Xanthine oxidase.
43-6. Formation of uric acid.
43-7. Purine catabolism.
43-8. Gout.
43-9. Allopurinol.
43-10. Secondary hyperuricemias.
43-11. Lesch-Nyhan syndrome.
43-12. Orotic aciduria.
43-13. Adenosine deaminase deficiency.
43-14. Give the enzyme defect in the following conditions: (a) Lesch-Nyhan syndrome; (b) Orotic aciduria.

CHAPTER 44: DNA STRUCTURE, REPLICATION

44-1. Watson and Crick model of DNA.
44-2. Base pairing rule.
44-3. Difference between DNA and RNA.
44-4. DNA polymerase.
44-5. Okazaki pieces.
44-6. Repair mechanisms of DNA.
44-7. Inhibitors of replication.
44-8. Antimetabolites.

CHAPTER 45: TRANSCRIPTION

45-1. RNA polymerase.
45-2. Reverse transcriptase.
45-3. Post-transcriptional modifications.
45-4. Introns and exons.
45-5. Inhibitors of transcription.

CHAPTER 46: TRANSLATION

46-1. Function of mRNA.
46-2. Ribosomes.

46-3. Initiation of translation.
46-4. Translocation.
46-5. Structure and function of tRNA.
46-6. Genetic code.
46-7. Degeneracy of codons.
46-8. Anticodon.
46-9. Post-translational modifications.
46-10. Inhibitors of protein biosynthesis.

CHAPTER 47: MUTATIONS AND GENE EXPRESSION

47-1. Induction.
47-2. Repression.
47-3. Operon concept.
47-4. Mutations.
47-5. Mutagens.
47-6. Point mutation.
47-7. Frame-shift mutation.
47-8. Conditional mutation.

CHAPTER 48: RECOMBINANT DNA TECHNOLOGY

48-1. Restriction endonucleases.
48-2. Gene therapy.

CHAPTER 49: MOLECULAR DIAGNOSTICS

49-1. Southern blotting.
49-2. Western blot technique.
49-3. Northern blotting
49-4. In situ hybridization
49.5. DNA finger printing
49-6. Restriction fragment length polymorphism (RFLP).
49-7. Polymerase chain reaction.
49-8. Monoclonal antibodies.
49-9. Hybridoma technology.
49-10. DNA sequencing.

CHAPTER 50: HORMONES, MECHANISM

50-1. G-proteins.
50-2. Transduction of message.
50-3. Second messenger.

CHAPTER 51: PITUITARY HORMONES

51-1. Gonadotropins.

CHAPTER 52: STEROID HORMONES

52-1. Biological effects of glucocorticoids.

CHAPTER 53: THYROID HORMONES

53-1. TSH stimulation test.
53-2. T3 suppression test.

CHAPTER 55: IMMUNOCHEMISTRY

55-1. Immunoglobulins.
55-2. Bence Jones proteins.
55-3. Multiple myeloma.
55-4. Immunoglobulin E and its clinical significance.

CHAPTER 57: BIOCHEMISTRY OF CANCER

57-1. Antimetabolites.
57-2. Oncogenes.
57-3. Oncosuppressor genes.
57-4. Tumor markers.
57-5. Alpha feto protein.
57-6. Carcinoembryonic antigen.
57-7. Name the tumor marker most appropriate for the following: (a) Prostate carcinoma; (b) Choriocarcinoma; (c) Colon cancer; (d) Hepatoma; (e) Pheochromocytoma; (f) Carcinoid syndrome; (g) Bone metastasis.

CHAPTER 58: TISSUE PROTEINS

58-1. Collagen
58-2. Prions.

CHAPTER 59: RADIOISOTOPES

59-1. Applications of isotopes for diagnosis
59-2. Biological effects of radiation on tissues.

Part 3: Reference Values, Recommended Daily Allowances and Deficient Enzymes

Table 3.1: Reference values

Students should remember the following important normal values. More details are given in the Textbook, Appendix.

Analyte (blood / serum / plasma)	*Value*
Alanine aminotransferase (ALT)	10–35 IU/L
Albumin	3.5–5 g/dL
Alkaline phosphatase (ALP)	40–125 IU/L
Aspartate aminotransferase (AST)	8–20 IU/L
Bicarbonate ($HCO3^-$)	22–26 mEq/L
Bilirubin, total	0.2–1 mg/dL
Calcium	9–11 mg/dL
Chloride	96–106 mEq/L
Cholesterol, Total	150–200 mg/dL
(HDL fraction) Male Female	30–60 mg/dL 35–75 mg/dL
(LDL fraction) 30–39 years	80–175 mg/dL
Creatinine	0.7–1.4 mg/dL
Globulins	2.5–3.5 g/dL
Glucose (Fasting)	70–110 mg/dL
Hemoglobin Male Female	14–16 g/dL 13–15 g/dL
Hb A1C (glycohemoglobin)	4–5.5% of total Hb
pH	7.4
Potassium	3.5–5 mEq/L
Proteins—total	6–8 g/dL
Sodium	136–145 mEq/L
Triglycerides (fasting), Male Female	50–200 mg/dL 40–150 mg/dL
Urea	20–40 mg/dL
Uric acid	3.5–7 mg/dL

Table 3.2: Interpretation of blood values

Analyte	*Increased in*	*Lowered in*
ALT	Liver disorders	
AST	Myocardial infarction	
Albumin		Cirrhosis, nephrotic malnutrition
ALP	Liver and bone diseases	
Bicarbonate	Alkalosis	Acidosis
Bilirubin, total	Jaundices	
do, conjugated	Obstructive jaundice	
do, unconjugated	Hemolytic jaundice	
Calcium	Hyperparathyroid	Tetany, rickets
Cholesterol, Total	Diabetes, Atherosclerosis Obstructive jaundice	
Creatinine	Renal failure	
Globulins	Multiple myeloma Liver diseases	
Glucose (Fasting)	Diabetes mellitus	
Hemoglobin	Polycythemia	Anemia
Hb A1C	Diabetes mellitus	
pH	Alkalosis	
	Acidosis	
Proteins—total	Multiple myeloma	Malnutrition
Urea	Kidney diseases	
Uric acid	Gout	

Table 3.3: RDA of important nutrients

(Values are for normal adult males; appropriate increased allowance will be given to children and pregnant mothers)

Nutrient	*Daily requirement*
Proteins	1 g / kg body weight
Vitamins	
Vitamin A	750 µg
Vitamin D	10 µg
Thiamine (B_1)	1 mg
Riboflavin (B_2)	1.5 mg
Niacin	20 mg
Pyridoxine (B_6)	2 mg
Folic acid	100 µg
Vitamin B_{12}	1 µg
Vitamin C	70 mg
Minerals	
Calcium	500 mg
Phosphorus	500 mg
Sodium	5 g
Potassium	3 g
Iron	20 mg
Copper	2 mg
Iodine	200 µg
Zinc	10 mg
Selenium	100 µg

Table 3.4: Inborn errors of metabolisms

(Deficient enzymes of important inborn errors of metabolisms are given for ready reference. For more details, see the chapters in the Textbook)

Disease	*Deficient enzyme*	*Ch.*
Acute intermittant porphyria	Uroporphyrinogen-I synthetase	22
Albinism	Tyrosinase	18
Alkaptonuria	Homogentisic acid oxidase	18
Argininosuccinic aciduria	Argininosuccinate lyase	15
Citrullinemia	Argininosuccinate synthetase	15
Cystathionuria	Cystathionase	16
Fructose intolerance	Fructose-1-P-aldolase	11
Galactosemia	Galactose-1-phosphate uridyl transferase	11
Gaucher's disease	Beta glucosidase	14
Homocystinuria	Cystathionine synthase	16
Lesch-Nyhan syndrome	HGPRTase	43
Maple syrup urine disease	Branched chain keto acid decarboxylase	17
Methyl malonic aciduria	Methyl malonyl CoA mutase	12
Niemann-Pick disease	Sphingomyelinase	14
Orotic aciduria	Orotidine-5-P decarboxylase	43
Phenylketonuria	Phenyl alanine hydroxylase	18
Tay-Sachs disease	Hexosaminidase A	14
von Gierke's disease	Glucose-6-phosphatase	10
Wilson's disease	Copper transport protein	39

Part 4: Viva Voce

CHAPTER 2: CELL

2-1. What is the function of Golgi complex?
Maturation and processing of nascent proteins, glycosylation of proteins, secretion newly synthesized proteins.

2-2. What is the function of endoplasmic reticulum?
Biosynthesis of proteins, drug metabolism, desaturation of fatty acids.

2-3. Where does protein synthesis takes place?
On the walls of endoplasmic reticulum and also in cytosol.

2-4. What are cathepsins?
They are intracellular proteolytic enzymes.

2-5. What is the function of lysosomes?
They are bags of hydrolytic enzymes that bring about degradation of macromolecules.

2-6. What is lysozyme?
It is an enzyme present in external secretions.

2-7. What is the function of mitochondria?
Generation of ATP.

2-8. What are the important metabolic events taking place in mitochondria?
TCA cycle, electron transport chain, beta oxidation of fatty acids and urea cycle (Table 2.2).

2-9. What are ecto-enzymes?
Enzymes seen on the outer part of the cell membrane.

2-10. Give examples of ecto-enzymes.
Alkaline phosphatase, 5'nucleotidase.

2-11. What is the structure of cell membrane?
Fluid mosaic model.

2-12. What are the characteristics of fluid mosaic model?
Membrane is composed of lipid bilayer. Phospholipids are arranged in bilayers with a hydrophobic core.

2-13. What are the components of membrane that alter the fluidity?
Cholesterol and unsaturated fatty acids.

2-14. What are the different types of transport mechanisms?
Passive and active. Passive type is subclassified as simple diffusion and facilitated diffusion.

2-15. What are salient features of facilitated diffusion?
It is carrier mediated. It does not require energy directly.

2-16. Give an example of facilitated transport.
Glucose transporters.

2-17. Give some examples of ion channels.
Ion channels for calcium, potassium and chloride.

2-18. What are ionophores?
They are transport antibiotics which increase the permeability of membrane to ions, e.g. valinomycin, gramicidin.

2-19. What are the salient features of active transport?
It requires transporters. It requires energy. Transport is generally unidirectional.

2-20. Give examples of active transport systems.
Sodium pump, calcium pump.

2-21. What is the importance of sodium pump?
Cell has low intracellular sodium; but concentration of potassium inside the cell is high; this is maintained by sodium pump.

2-22. What is the clinical significance of sodium pump?
Digoxin increases the contractility of the cardiac muscle, by inhibiting the sodium pump.

2-23. What is a uniport?
It carries single solute across the membrane.

2-24. Give examples of uniport system.
Glucose transporter (GluT2) operating in most of the cells is an example. Calcium pump is another example.

2-25. What is co-transport?
If transfer of one molecule depends on simultaneous transfer of another molecule, it is called co-transport system.

2-26. What is symport?
In symport, the transporter carries two solutes in the same direction across the membrane.

2-27. Give examples of symport.
Sodium dependent glucose transporter (S GluT). Amino acid transport is another example for symport.

2-28. What is antiport system?
The antiport system carries two solutes or ions in opposite direction.

2-29. Give examples of antiport.
Sodium pump or chloride-bicarbonate exchange in RBC.

CHAPTER 3: AMINO ACIDS, STRUCTURE

3-1. What are branched chain amino acids?
Valine, leucine and isoleucine (Fig 3.3).

3-2. What are hydroxy amino acids?
Serine and threonine (Fig. 3.4).

3-3. Name the sulfur-containing amino acids.
Cysteine and methionine (Fig. 3.5).

3-4. Name the acidic amino acids.
Aspartic acid and glutamic acid (Fig. 3.7).

3-5. What are the basic amino acids?
Lysine and arginine (Fig. 3.8).

3-6. Which amino acid contains thio-ether bond?
Methionine (Table 3.1).

3-7. Which amino acids have hydrophobic side chains?
Valine, leucine, isoleucine.

3-8. Give the names of aromatic amino acids.
Phenylalanine and tyrosine (Fig. 3.9).

3-9. What are heterocyclic amino acids?
Tryptophan and histidine (Figs 3.10 and 3.11).

3-10. Give an example of an imino acid.
Proline (Fig. 3.11).

3-11. Give examples of derived amino acids.
Hydroxyproline, hydroxylysine, ornithine, citrulline, homocysteine.

3-12. Arginine contains which special group?
Guanidinium group (Table 3.1).

3-13. Benzene group is present in which amino acid?
Phenylalanine (Table 3.1).

3-14. Phenol group is present in which amino acid?
Tyrosine (Table 3.1).

3-15. Tryptophan contains what special group?
Indole group (Table 3.1).

3-16. Which special group is present in Histidine?
Imidazole group (Table 3.1).

3-17. Name some hydrophobic amino acids.
Valine, leucine and isoleucine.

3-18. Pyrrolidine is present in which amino acid?
Proline (Table 3.1).

3-19. Hydrophobic bonds are formed in protein between which amino acids?
Valine, leucine and isoleucine residues.

3-20. What is the basis of classification of amino acids into ketogenic and glucogenic?
Ketogenic amino acids will enter into the metabolic pathway of fats, while glucogenic amino acids enter the pathway of glucose metabolism.

3-21. Name a purely ketogenic amino acid.
Leucine.

3-22. Name some glucogenic amino acids
Glycine; serine; aspartic acid.

3-23. What are essential amino acids?
They can not be synthesized in the body; and so, they are to be provided in the diet.

3-24. How many amino acids are essential?
Eight amino acids are essential; two are semi-essential and the rest 10 are non-essential (Box 3.1).

3-25. Name any three essential amino acids.
Isoleucine, leucine, threonine (Box 3.1).

3-26. Is phenylalanine an essential amino acid?
Yes (Box 3.1).

3-27. What about tyrosine?
Tyrosine is non-essential (Box 3.1).

3-28. What is iso-electric point?
The pH at which the molecule carries no net charge is called iso-electric point.

3-29. What are the characteristic features of iso-electric pH?
At iso-electric point the amino acid will carry no net charge; there is no mobility in electrical field.

Solubility will be minimum. The tendency for precipitation will be maximum.

3-30. Which amino acid has maximum buffering capacity at physiological pH?
Histidine.

3-31. Which amino acid is optically inactive?
Glycine.

3-32. What are the isomeric forms of amino acids?
D and L varieties.

3-33. What are natural amino acids?
Only L amino acids are seen in nature.

3-34. What is decarboxylation of an amino acid?
The carboxyl group is removed from the amino acids to form the corresponding amine.

3-35. Give examples of decarboxylation reactions.
Histidine to histamine; tyrosine to tyramine; tryptophan to tryptamine.

3-36. What is produced when glutamic acid is decarboxylated?
Gamma amino butyric acid or GABA.

3-37. What is glutamine?
That is the amide of glutamic acid.

3-38. What is an amide?
The extra carboxyl group (other than alpha carboxyl) can combine with ammonia to form the corresponding amide.

3-39. What is transamination?
The alpha amino group of amino acid can be transferred to alpha keto acid to form the corresponding new amino acid and alpha keto acid (Fig. 3.18).

3-40. Give an example of transamination reaction.
Glutamic acid + pyruvic acid to alpha keto glutarate + alanine (Fig. 3.18).

3-41. What is the biological significance of transamination reaction?
These are important for the inter-conversion of amino acids. Non-essential amino acids are synthesized by this process.

3-42. What is the clinical significance of transaminases?
Transaminases in blood are elevated in liver and heart diseases.

3-43. What is the significance of SH groups in proteins?
The SH group of cysteine can form a disulfide (S-S) bond with another cysteine residue. The two cysteine residues can connect two polypeptide chains by the formation of interchain disulfide bonds (Fig. 3.20).

3-44. Phosphorylation is taking place on which amino acid?
Serine and threonine.

CHAPTER 4: PROTEINS, STRUCTURE

4-1. How proteins are made up of?
Proteins are made by polymerization of amino acids through peptide bonds (Fig. 4.1A).

4-2. What is a peptide bond?
Alpha carboxyl group of one amino acid reacts with alpha amino group of another amino acid to form a peptide bond.

4-3. What is a dipeptide?
Two amino acids are combined to form a dipeptide.

4-4. How many peptide bonds are present in a tripeptide?
A tripeptide is a combination of three amino acids; so there are two peptide bonds.

4-5. What is a polypeptide?
A combination of 10 to 50 amino acids is called as a polypeptide.

4-6. What are the levels of organizations of proteins?
Proteins have primary, secondary, tertiary and quaternary levels of organization.

4-7. What is meant by primary structure of a protein?
It denotes the number and sequence of amino acids in the protein.

4-8. What force maintains the primary structure?
The primary structure is maintained by the covalent bonds of the peptide linkages.

4-9. How are the end amino acids of proteins called?
The end where there is a free alpha amino group, is called the amino terminal (N-terminal) end. The other end of the polypeptide chain is called the carboxy terminal end (C-terminal), where there is a free alpha carboxyl group.

4-10. What is a pseudopeptide?
The pseudopeptide is a peptide bond formed by carboxyl group, other than that of alpha position.

4-11. Can you give an example of a pseudopeptide?
Glutathione (gamma-glutamyl-cysteinyl-glycine).

4-12. What are the salient structural features of insulin?
It has two polypeptide chains with 51 amino acids. Chain A has 21 amino acids and Chain B has 30 amino acids. The two chains are held together by disulfide bridges (Fig. 4.4).

4-13. What is pro-insulin?
Insulin is synthesized by the beta cells of pancreas as a single polypeptide chain with 86 amino acids. The middle part is removed as C peptide; the remaining part becomes A and B chains (Fig. 4.5).

4-14. What is a mutation?
Amino acid change in the linear sequence is called a mutation.

4-15. Can you give an example?
Sickle cell anemia due to hemoglobin S (HbS).

4-16. What are the forces that maintain the secondary, tertiary and quaternary structures of a protein?
Hydrogen bonds; electrostatic bonds; van der Waal's forces and hydrophobic bonds.

4-17. What are the salient features of alpha structure of proteins?
It is a right handed spiral structure. Each turn is formed by 3.6 amino acid residues.

4-18. What is secondary structure of a protein?
Secondary structure denotes the configurational relationship between residues, which are about 3–4 amino acids apart. In other words, secondary level defines the organization at immediate vicinity of amino acids.

4-19. What is meant by tertiary structure of a protein?
The tertiary structure denotes three-dimensional structure of the whole protein. It defines the steric relationship of amino acids which are far apart from each other in the linear sequence.

4-20. What is quaternary structure of a protein?
Certain polypeptides will aggregate to form one functional protein. This is referred to as the quaternary structure.

4-21. Name proteins having quaternary structure.
Hemoglobin; immunoglobulins.

4-22. What are the reagents that are used for identifying the first amino acid in a protein?
Fluorodinitrobenzene; dansyl chloride; phenylisothiocyanate.

4-23. What is iso-electric point of a protein?
At the iso-electric point, the number of anions and cations present on the protein molecule will be equal and the net charge is zero.

4-24. What are the features of iso-electric point?
At the pI value, the proteins will not migrate in an electrical field; solubility, buffering capacity and viscosity will be minimum and precipitation will be maximum.

4-25. What is the iso-electric pH of human albumin?
It is 4.7.

4-26. How proteins are precipitated from solution?
Any factor which neutralizes the charge or removes water of hydration will cause precipitation of proteins.

4-27. How is albumin precipitated?
By full saturation of ammonium sulfate.

4-28. What about gobulins?
Globulins are precipitated by half saturation of ammonium sulfate.

4-29. Give example of iso-electric precipitation.
Casein is precipitated when the solution is brought to iso-electric pH.

4-30. What is the iso-electric pH of casein?
It is 4.6.

4-31. What are the features of denaturation?
The secondary, tertiary and quaternary structures are lost; but primary structure is preserved. The functional activity is lost. Denatured proteins are insoluble and easily precipitated.

4-32. What is heat coagulation?

When heated at iso-electric point, some proteins will denature irreversibly to produce thick floating conglomerates called coagulum. This is called heat coagulation.

4-33. Give examples of proteins that coagulate.

Albumin is easily coagulated, and globulins to a lesser extent.

4-34. How are proteins classified?

They may be classified (a) depending on the function, (b) based on the physicochemical characteristics, or (c) based on their nutritional value.

4-35. How are proteins classified on physical basis?

Simple proteins, conjugated proteins and derived proteins.

4-36. Give examples of simple proteins.

Albumins, globulins, protamines, prolamines, lectins, scleroproteins.

4-37. Give examples of scleroproteins.

Collagen of bone, cartilage and tendon; keratin of hair.

4.38. What are conjugated proteins?

Combinations of protein with a non-protein part, called prosthetic group.

4-39. How are conjugated group subclassified?

Glycoproteins, lipoproteins, nucleoproteins, chromoproteins, phospho-proteins and metalloproteins.

4-40. Give some examples of chromoproteins

Hemoglobin; flavoproteins, visual purple.

4-41. Give an example of a nutritionally rich protein (first class protein).

Casein.

4-42. Why some proteins are nutritionally poor?

They lack in many essential amino acids and a diet based on these proteins will not even sustain the body weight.

4-43. Give an example of nutritionally poor protein.

Zein from corn lacks tryptophan and lysine.

4-44. What is the advantage of biuret method?

The biuret method is the simple one step process, and is the most widely used method for plasma protein estimations.

4-45. What is the disadvantage of biuret method?
The sensitivity of the method is less and is unsuitable for estimation of proteins in milligram or microgram quantities.

CHAPTER 5: ENZYMES, GENERAL CONCEPTS

5-1. How are enzymes classified?
They are classified into 5 major classes (Box 5.1).

5-2. What are those classes?
Oxidoreductases, transferases, hydrolases, lyases, isomerases and ligases (Box 5.1).

5-3. Give an example of oxidoreducatase.
Alcohol dehydrogenase (Box 5.1).

5-4. What is the function of transferases?
Transfer of groups other than hydrogen (Box 5.1).

5-5. Give an example of transferase.
Hexokinase (Box 5.1).

5-6. What is the function of hydrolases?
Cleave the bond after adding water (Box 5.1).

5-7. Give an example of a hydrolase
Acetylcholinesterase (Box 5.1).

5-8. Peptidases are classified under which class of enzyme?
Hydrolases (Box 5.1).

5-9. What is the function of lyases?
Cleave bond without adding water (Box 5.1).

5-10. Which enzyme will add water to a double bond, without breaking the bond?
Hydratase (Box 5.1).

5-11. What is the function of ligases?
ATP dependent condensation of two molecules.

5-12. What is the difference between synthase and synthetase?
Synthetases are ATP-dependent enzymes, Synthases do not require ATP directly; they belong to classes other than ligases (Box 5.2).

5-13. Give examples of synthetases.
Carbamoyl phosphate synthetase; argininosuccinate synthetase (Box 5.2).

5-14. Give examples of synthases.
Glycogen synthase, ALA synthase, IMP synthase.

5-15. What are co-enzymes?
They are non-protein part of enzyme. They are low molecular weight organic substances, without which the enzyme cannot exhibit any reaction. Co-enzyme accepts one of the products of the reaction; and so act as a co-substrate.

5-16. How are co-enzymes classified?
(a) Those taking part in reactions catalyzed by oxidoreductases by donating or accepting hydrogen atoms or electrons. (b) Those co-enzymes taking part in reactions transferring groups other than hydrogen.

5-17. Give some examples of co-enzymes involved in oxidoreductases.
NAD^+, $NADP^+$, FAD.

5-18. What is the full form of NAD?
Nicotinamide adenine dinucleotide.

5-19. What is the full form of ATP?
Adenosine triphosphate (Fig. 5.3).

5-20. What is the function of ATP?
It is the energy currency in the body. During the oxidation of foodstuffs, energy is released, a part of which is stored as chemical energy in the form of ATP.

5-21. Name some enzymes containing copper.
Superoxide dismutase; tyrosinase; cytochrome oxidase.

5-22. Which metal is required for the kinases?
Magnesium (Table 5.2).

5-23. Which enzyme contains molybdenum?
Xanthine oxidase (Table 5.2).

5-24. Name some iron containing enzymes.
Cytochrome oxidase, catalase, peroxidase (Table 5.2).

5-25. What is Michaelis-Menten theory ?
It is otherwise called enzyme-substrate complex theory. The enzyme combines with the substrate, to form an enzyme-substrate complex, which immediately breaks down to the enzyme and the product.

5-26. What is Fischer's theory?
It states that the three-dimensional structure of the active site of the enzyme is complementary to the substrate. Enzyme and substrate fit like a key and its lock.

5-27. What is Koshland's induced fit theory?
Substrate induces conformational changes in the enzyme, such that precise orientation of catalytic groups is effected.

5-28. What is the active site of an enzyme?
That area of the enzyme where catalysis occurs is referred to as active site or active center.

5-29. What is meant by serine proteases?
Proteases (proteolytic enzymes) having a serine residue at its active center.

5-30. Give an example of a serine protease.
Trypsin, chymotrypsin, thrombin (Table 5.3).

5-31. What do enzymes on activation energy?
Enzymes lower activation energy (Fig. 5.4).

5-32. What is the action of enzymes on equilibrium?
They increase the chemical reaction, but do not alter equilibrium of the reaction.

5-33. What are factors influencing enzyme reaction?
Enzyme concentration, substrate concentration, product concentration, temperature, pH and presence of activators or inhibitors.

5-34. What is Km value?
Substrate concentration (expressed in mole/L) at half-maximal velocity is the Km value.

5-35. What does it indicate?
It denotes that 50% of enzyme molecules are bound with substrate molecules at that particular substrate concentration.

5-36. What is the significance of Km value?
Km is independent of enzyme concentration. Km value is thus a constant for an enzyme. It is the characteristic feature of a particular enzyme for a specific substrate.

5-37. What is the relation of Km value with the affinity?
Km denotes the affinity of enzyme to substrate. Thus, the lesser the numerical value of Km, the affinity of the enzyme for the substrate is more.

5-38. What is the effect of temperature on enzyme velocity?
The velocity of reaction increases when temperature is increased; reaches a maximum and then falls (Bell-shaped curve) (Fig. 5.17).

5-39. Why it falls?
When temperature is more than 50°C, heat denaturation occurs with consequent loss of tertiary structure of protein.

5-40. What is effect of pH on activity of an enzyme?
Each enzyme has an optimum pH, on both sides of which the velocity will be drastically reduced (Fig. 5.18).

5-41. What is the explanation for the effect of pH?
The pH decides the charge on the amino acid residues at the active site. The net charge on the enzyme protein would influence substrate binding and catalytic activity.

5-42. What is the optimum pH of usual enzymes?
Usually, enzymes have the optimum pH between 6 and 8.

5-43. Give some exceptions for this general rule.
Pepsin (optimum pH 1–2); Alkaline phosphatase (optimum pH 9–10) and Acid phosphatase (4–5) (Fig. 5.18).

5-44. What is a zymogen?
It is otherwise called pro-enzyme. Inactive zymogen is activated by removal of a piece of the pro-enzyme.

5-45. Give an example of zymogen activation.
By removing a small polypeptide from trypsinogen, the active trypsin is formed.

5-46. What is the significance of zymogen activation?
Gastrointestinal enzymes are synthesized in the form of pro-enzymes, and only after secretion into the alimentary canal, they are activated. This prevents autolysis of cellular structural proteins.

5-47. What are the different types of inhibitions of enzyme?
Competitive inhibition; non-competitive inhibition; suicide inhibition; and allosteric regulation.

5-48. What are the features of competitive inhibition?
1. Competitive inhibitor is a structural analog. 2. It is reversible. 3. Km is increased. 4. Vmax is not changed.

5-49. Give examples of competitive inhibition.
Malonate inhibits succinate dehydrogenase.

5-50. Give examples of clinical application of competitive inhibition.

(a) Sulfonamide inhibits PABA incorporation in bacteria; and so acts as an antibacterial agent. (b) Methotrexate inhibits folate reductase. (c) Dicoumarol inhibits vitamin K.

5-51. What are the features of non-competitive inhibition?

1. Non-competitive inhibitor has no structural similarity with the substrate. 2. It is generally not reversible. 3. Km is not changed. 4. Vmax is reduced (Fig. 5.23).

5-52. Give examples of non-competitive inhibition.

(a) Di-isopropyl fluorophosphate inhibits trypsin. (b) Fluoride inhibits enolase.

5-53. What is suicide inhibition?

In suicide inhibition, the structural analog is converted to a more effective inhibitor with the help of the enzyme to be inhibited. The inhibitor makes use of the enzyme's own reaction mechanism to inactivate it.

5-54. What is another term for suicide inhibition?

Mechanism based inactivation.

5-55. Give examples for suicide inhibition.

(a) Ornithine decarboxylase (ODC) is inhibited by difluromethyl ornithine (DFMO). (b) Allopurinol is oxidized by xanthine oxidase to alloxanthine that is a strong inhibitor of xanthine oxidase.

5-56. What is allosteric inhibition?

Allosteric enzyme has one catalytic site where the substrate binds and a separate allosteric site where the modifier binds.

5-57. What are the features of allosteric inhibition?

(1) The inhibitor is not a substrate analog. (2) It is partially reversible, when excess substrate is added. (3) Most allosteric enzymes are made up of subunits.

5-58. Give examples for allosteric inhibition.

ALA synthase, aspartyl trans carbamoylase, HMG CoA reductase (Table 5.7).

5-59. What is covalent modification?
Addition or removal of a group to the enzyme by a covalent bond.

5-60. Give some examples of covalent modification.
Glycogen phosphorylase is active when phosphorylated and inactive when dephosphorylated.

5-61. What is meant by induction?
Induction is effected at the level of DNA. The inducer will relieve the repression on the operator site and will remove the block on the biosynthesis of the enzyme molecules.

5-62. Give an example of induction.
Induction of lactose-utilizing enzymes in the bacteria when the media contains lactose in the absence of glucose. In humans, ALA synthase is induced by barbiturates.

5-63. What are constitutive enzymes?
Enzymes whose concentration in a cell is always the same.

5-64. What is repression?
Repression acts at the gene level; the number of enzyme molecules is reduced in the presence of repressor.

5-65. Give an example of repression.
ALA synthase is repressed by the heme.

5-66. What are the types of specificity?
Absolute specificity, group specificity and streo-specificity.

5-67. Give an example for absolute specificity.
Urea is the only substrate for urease.

5-68. Give an example for group specificity.
Trypsin can hydrolyze peptide bonds formed by carboxyl groups of arginine or lysine residues.

5-69. What are iso-enzymes?
They are physically distinct forms of the same enzyme activity. They have identical catalytic properties, but differ in structure.

CHAPTER 6: CARBOHYDRATES, CHEMISTRY

6-1. How carbohydrates are classified?
Based on the number of the sugar units available, they are classified as monosaccharides, disaccharides, oligosaccharides and polysaccharides.

6-2. What is a monosaccharide?
Molecules having only one actual or potential sugar group are called monosaccharides.

6-3. What is a polysaccharide?
They contain more than 10 sugar units.

6-4. How are they combined together?
Through glycosidic linkages.

6-5. How are monosaccharides further classified?
Sugars having aldehyde group are called aldoses and sugars with keto group are ketoses.

6-6. Name some important monosaccharides.
Glucose, fructose, mannose, galactose (Fig. 6.4).

6-7. What are pentoses?
Monosaccharides with 5 carbon atoms.

6-7. Name a few pentoses.
Arabinose, xylose, ribose.

6-8. Which is the reference carbon atom in sugars?
Penultimate carbon atom (Fig. 6.3).

6-9. What is the difference between D and L sugars?
They are mirror images with reference to penultimate carbon atom (Fig. 6.3).

6-10. Which isomer is common in nature?
D variety of sugars are common in nature.

6-11. What is the difference between glucose and galactose?
Galactose is the 4th epimer of glucose (Fig. 6.4).

6-12. Galactose is present in which food?
Lactose is present in milk. Lactose contains galactose and glucose.

6-13. What is epimerism?
When sugars are different from one another, only in configuration with regard to a single carbon atom (other than the reference carbon atom), they are called epimers.

6-14. Give an example.
Glucose and mannose are an epimeric pair which differ only with respect to carbon atom 2. Similarly, galactose is the 4th epimer of glucose.

6-15. Anomerism is produced with reference with which carbon atom?
With reference to the first carbon atom in aldoses and second carbon atom in ketoses.

6-16. How alpha and beta sugars are produced?
These are anomers (Fig. 6.5).

6-17. What is the basis of mutarotation?
It is due to the anomeric carbon atom.

6-18. What is the difference between glucose and fructose?
Glucose is aldohexose; fructose is a ketohexose.

6-19. What is the principle of Benedict's test?
In alkaline medium, sugar reduces cupric ions, red precipitate (Fig. 6.9).

6-20. Benedict's test is commonly done for what?
To detect the presence of glucose in urine.

6-21. Name a few reducing sugars.
Glucose, fructose, mannose.

6-22. Keto group is non-reducing; but fructose reduces Benedict's solution; what is the cause for this anomaly?
In alkaline medium, ketone group is converted to aldehyde, through enediol formation (Fig. 6.10).

6-23. In the case of sugars, which of the properties go hand in hand?
Reducing property; osazone formation and mutarotation.

6-24. Glucose and fructose will form identical osazones, why?
The difference in glucose and fructose is dependent on the first and second carbon atoms; and this is masked by the osazone formation.

6-25. On oxidation of glucose, what are produced?
Glucuronic acid; gluconic acid and glucosaccharic acid.

6-26. Name some important disaccharides.
Sucrose, lactose, maltose (Box 6.3).

6-27. What is the glycosidic linkage in lactose?
Beta 1-4 linkage (Fig. 6.20).

6-28. What is the glycosidic linkage in sucrose?
1-2 linkage (Fig. 6.19A).

6-29. Which disaccharide has no free aldehyde or ketone group?
Sucrose (Fig. 6.19A).

6-30. Sucrose (containing glucose and fructose) is a non-reducing sugar, why?
Because the glycosidic linkage in sucrose involves 1st carbon of glucose and 2nd carbon of fructose; so both reducing groups are masked.

6-31. Hydrolysis of maltose will give rise to what?
Two glucose units.

6-32. Which is the sugar found in milk?
Lactose.

6-33. What are the component monosaccharides of lactose?
Galactose and glucose (Fig. 6.20).

6-34. Sucrose consists of what monosaccharides?
Glucose + fructose (Fig. 6.19).

6-35. Name reducing disaccharides.
Lactose and maltose (Box 7.3).

6-36. How polysaccharides are classified?
Homopolysaccharides (homoglycans) and heteropolysaccharides (heteroglycans).

6-37. What is a homoglycan?
They are composed of single kind of monosaccharides.

6-38. Give examples of homopolysaccharides.
Starch, glycogen.

6-39. What are heteropolysaccharides?
They are composed of two or more different monosaccharides.

6-40. What are the characteristics of glycogen ?
It is composed of glucose units. It is the stored form of carbohydrate in animal kingdom. It has a highly branched structure.

6-41. Give examples of heteropolysaccharides.
Agar, hyaluronic acid, heparin, chondroitin sulfate.

6-42. What are mucopolysaccharides?
They contain uronic acid and amino sugars.

6-43. Which heteropolysaccharide does not contain uronic acid?
Keratan sulfate.

6-44. Hyaluronic acid is seen in which tissues?
Connective tissue, synovial fluid, tendons, vitreous humor.

CHAPTER 7: LIPIDS, CHEMISTRY

7-1. How are lipids classified?
Simple, compound and derived.

7-2. Classify fatty acids.
Depending on the total number of carbon atoms, they are classified as even chain and odd chain.

7-3. Which type is prevalent in human body?
Even chain fatty acids.

7-4. Name some saturated fatty acids.
Palmitic acid, stearic acid (Table 7.2).

7-5. Name some unsaturated fatty acids.
Oleic, linoleic, linolenic and arachidonic acids (Table 7.2).

7-6. Name some polyunsaturated fatty acids.
Linoleic, linolenic and arachidonic acids (Fig. 7.2).

7-7. How many double bonds are there in arachidonic acid?
Four double bonds (Fig. 7.2).

7-8. Which contains good quantity of PUFA?
Vegetable oils such as sunflower, ground nut oil (Table 7.4).

7-9. Which contains very low level of PUFA?
Animal fats (Table 7.4).

7-10. Which fatty acids are common in human fat?
Mainly Oleic acid; then comes palmitic acid and linoleic acid.

7-11. What is the difference between fats and oils?
Fats are solids, oils are liquids at room temperature.

7-12. What is the advantage of storing energy as triglycerides in the body?
(a) Space requirement is less as storage does not require water; (b) can be mobilized whenever required; (c) capacity for storage is unlimited.

7-13. What is saponification?
Hydrolysis of fat by alkali is called as saponification.

7-14. What is the significance of iodine value?
It gives the idea of how much double bonds are there in oils.

7-15. What is lecithin?
Phosphatidylcholine (Fig. 7.8).

7-16. What is phosphatidic acid?
It is made up of one glycerol, two fatty acid residues and a phosphoric acid (Fig. 7.6).

7-17. What is cephalin?
Phosphatidylethanolamine (Fig. 7.9).

7-18. What is cardiolipin?
Diphosphatidylglycerol.

7-19. What is nitrogenous base present in lecithin?
Choline (Fig. 7.8).

7-20. What is sphingomyelin?
Sphingosine is attached to a fatty acid to form a ceramide. Ceramide with choline is sphingomyelin (Fig. 7.14).

7-21. Spingomyelin, on hydrolysis produces what?
Sphingosine, fatty acid, phosphoric acid, choline.

7-22. What is the special feature of sphingomyelin?
Sphingomyelins are the only sphingolipid that contain phosphate and have no sugar moiety.

7-23. Cerebroside contains what?
Sphingosine, fatty acid, hexose.

7-24. A ganglioside on hydrolysis, will give rise to?
Sphingosine, fatty acid, N-acetylneuraminic acid.

CHAPTER 8: OVERVIEW OF METABOLISM

8-1. In normal resting state, most of the glucose is utilized by which tissue?
Brain.

8-2. During starvation, which substances are increased in blood?
Ketone bodies, glucagon, epinephrine.

8-3. Name organs which cannot utilize fatty acid.
Brain and RBCs.

CHAPTER 9: MAJOR METABOLIC PATHWAYS OF GLUCOSE

9-1. Glucose is absorbed at the luminal side of gastro intestinal cells by which mechanism?
Carrier mediated co-transport with sodium, named as sodium dependent glucose transporter (SGluT) (Fig. 9.1).

9-2. How glucose is released from intestinal cells into the bloodstream?
Glucose transporter type 2 (GluT2) (Fig. 9.2A).

9-3. How glucose is taken up by cells?
In tissues GluT2 helps in absorption of glucose from blood.

9-4. What is the importance of GluT4?
It is the glucose transporter present in muscle and adipose tissue.

9-5. What is its clinical significance?
Insulin induces these transporters. In diabetes mellitus, entry of glucose into muscle is decreased, because GluT4 is reduced.

9-6. What is glycolysis?
In this pathway, glucose is converted to pyruvate or lactate, along with production of a small quantity of energy.

9-7. In which condition pyruvate is produced?
In aerobic condition pyruvate is produced.

9-8. In which condition lactate is produced?
When oxygen is lacking, lactate is produced.

9-9. What is the significance of glycolysis?
Anaerobic glycolysis forms the major source of energy in actively contracting muscles. Moreover, glycolysis is the only source of energy in RBCs.

9-10. What is the difference between glucokinase and hexokinase?
Both add phosphate group to glucose. But glucokinase is present only in liver; acts specifically on glucose, and is active when glucose level in blood is increased after a food (Table 9.2).

9-11. Which tissues prefer anaerobic glycolysis?
RBCs, exercising muscle, and cancer cells

9-12. During glycolysis, energy is produced during which steps?
Step 5, Glyceraldehyde-3-phosphate to 1,3-bisphosphoglycerate; Step 6, 1,3-bisphospho- glycerate to 3-phosphoglycerate; and Step 9, Phosphoenol pyruvate to pyruvate (Table 9.5).

9-13. Fluoride ions inhibit which enzyme?
Enolase (Step 8).

9-14. What is the clinical significance of the above inhibition?
Fluoride is used to prevent glycolysis, as preservative for blood before glucose estimation.

9-15. In glycolysis, NAD^+ is reduced to NADH in which reaction ?
Glyceraldehyde-3-phosphate dehydrogenase reaction (Fig. 9.4).

9-16. What are substrate level phosphorylation steps in glycolysis?
1,3-bisphosphoglycerate kinase (step 6) and pyruvate kinase (step 9) (Fig. 9.4).

9-17. What is the purpose of lactic acid production under anaerobic conditions?
NADH generated in the 5th step has to be oxidised to NAD^+. This can be done by oxygen. But when oxygen is lacking, the 5th step has to be coupled with the 10th step for regeneration of NAD (See Fig. 9.15).

9-18. What is Cori's cycle (or lactic acid cycle)?
During exercise, lactate is produced in muscle. This lactate diffuses into the blood. Lactate then reaches liver, where it is oxidized to pyruvate. It is then becomes glucose. This glucose can enter into blood and then taken to muscle (Fig. 9.19).

9-19. What is the purpose of Cori's cycle?
By this means, the lactate is efficiently reutilized.

9-20. What are the key glycolytic enzymes?
Glucokinase; phosphofructokinase; pyruvate kinase (Fig. 9.4).

9-21. What is the action of insulin on glycolysis?
Insulin stimulates glycolysis.

9-22. What is the energy yield from glycolysis during anaerobic conditions?
2 ATP (Table 9.4).

9-23. In aerobic glycolysis, the net yield from one glucose molecule is how much?
7 ATP (Table 9.5).

9-24. During complete oxidation, what is the net yield of ATP from one glucose molecule?
32 ATP (Table 9.6).

9-25. How many ATPs are generated per one rotation of the citric acid cycle?
10 ATP (Table 9.6).

9-26. What is function of 2,3-bisphosphoglycerate?
It reduces the affinity of hemoglobin towards oxygen.

9-27. What are the steps in which carbon dioxide is released from a glucose molecule?
Pyruvate dehydrogenase; isocitrate dehydrogenase; alpha keto glutarate dehydrogenase.

9-28. What are the co-enzymes necessary for oxidative decarboxylation of pyruvate?
Thiamine pyrophosphate; NAD; FAD; lipoic acid; co-enzyme A (Fig. 9.22).

9-29. There is no net synthesis of glucose from fatty acids, why?
Pyruvate to acetyl CoA is a totally irreversible reaction.

9-30. Which enzyme irreversibly channels glucose to energy production ?
Pyruvate dehydrogenase is the irreversible reaction.

9-31. What is gluconeogenesis?
Production of glucose from non-carbohydrate sources.

9-32. What are those non-carbohydrate sources? (What are the substrates for gluconeogenesis?)
Glucogenic amino acids and lactate (Fig. 9.27).

9.33. What are the key gluconeogenic enzymes?
Pyruvate carboxylase; phosphoenol pyruvate carboxy kinase; fructose-1,6-bisphosphatase and glucose-6-phosphatase (Fig. 9.27).

9.34. Malate shuttle is used for what purpose?
Reactions of gluconeogenesis are taking place in cytosol. Hence the oxaloacetate has to be transported from mitochondria to cytosol. This is done by malate shuttle (Fig. 9.25).

9-35. Gluconeogenesis is seen in which tissue?
Liver.

9-36. How many ATP molecules are required to convert 2 molecules of pyruvate into glucose?
Six.

9-37. Blood glucose level can be raised by gluconeogenesis only by liver, why?
Glucose-6-phosphatase is present only in liver.

9-38. Muscle glycogen will not serve as a precursor of blood sugar, why?
Glucose-6-phosphatase is absent in muscle.

9-39. Which amino acids are both ketogenic and glucogenic?
Tyrosine and tryptophan (Fig. 9.27).

9-40. What hormone will inhibit gluconeogenesis?
Insulin (Fig. 9.33).

9-41. What hormones will stimulate gluconeogenesis?
Glucagon and glucocorticoids (Fig. 9.33).

9-42. What is the significance of gluconeogenesis?
Gluconeogenesis is necessary to maintain blood glucose level, especially under conditions of starvation.

CHAPTER 10: OTHER PATHWAYS OF GLUCOSE

10-1. What is the key enzyme of glycogenolysis?
Glycogen phosphorylase.

10-2. Which hormones enhance glycogenolysis?
Adrenaline and glucagon causes glycogenolysis.

10-3. What is the mechanism of action of adrenaline?
Adrenaline increases cyclic AMP level which activates glycogen phosphorylase (Fig. 10.5).

10-4. What will activate glycogen phosphorylase?
Epinephrine, glucagon, cyclic AMP.

10-5. What is cyclic AMP (cAMP)?
It is a second messenger (Fig. 10.5).

10-6. How cyclic AMP is produced?
Adenyl cyclase enzyme acts on ATP to produce cAMP.

10-7. Which is the defective enzyme in von Gierke's disease (glycogen storage disease type I)?
Glucose-6-phosphatase (Table 10.2).

10-8. What is the characteristic clinical features of von Gierke's disease?
Fasting hypoglycemia, which does not respond to adrenaline is very characteristic.

10-9. Which enzyme generates NADPH?
Glucose- 6-phosphate dehydrogenase (Fig. 10.8).

10-10. Which is the key enzyme of hexose monophosphate shunt pathway?
Glucose-6-phosphate dehydrogenase.

10-11. What are the tissues in which HMP shunt pathway is significant?
Liver, adipose tissue, RBC, adrenal cortex, ovary, testis, mammary gland, lens.

10-12. What hormone controls HMP shunt pathway?
Insulin stimulates the pathway by activating the key enzyme.

10-13. What is the purpose of HMP shunt pathway?
It generates NADPH.

10-14. What is the use of NADPH in biological systems?
For reductive biosynthesis.

10-15. Which synthetic pathways need NADPH?
Fatty acid biosynthesis; synthesis of cholesterol, steroid hormones.

10-16. Apart from reductive synthesis, NADPH is used for what purpose?
It is necessary to keep the integrity of RBC membrane; for keeping transparency of lens; and for superoxide production inside macrophages.

10-17. Is NADPH used for ATP generation?
No. NADPH is not used for ATP generation.

10-18. What is the manifestation of glucose-6-phosphate dehydrogenase deficiency?
Drug induced hemolysis.

10-19. Acute hemolytic episode after administration of antimalarial drug is due to what?
Deficiency of glucose-6-phosphate dehydrogenase.

10-20. What is the most common enzyme deficiency?
Glucose-6-phosphate dehydrogenase deficiency.

10-21. What is clinical significance of transketolase?
The transketolase available in RBCs is an index of the thiamine status of an individual.

10-22. When is transketolase activity decreased?
Deficiency of Thiamine pyrophosphate (TPP).

10-23. What is the purpose of uronic acid pathway?
It is used for conjugation of bilirubin; steroids, synthesis of glucosaminoglycans.

10-24. In lower animals, uronic acid pathway is used for what ?
For synthesis of ascorbic acid (vitamin C) (Fig. 10.7).

10-25. What is essential pentosuria?
Excretion of pentose (L-xylulose) in urine.

10-26. What is it due to?
Due to the deficiency of xylulose reductase or xylulose reductase.

CHAPTER 11: OTHER PATHWAYS OF CARBOHYDRATES

11-1. What is fructose intolerance?
Due defective aldolase-B. So, fructose-1-phosphate accumulates.

11-2. What is the clinical application of fructose estimation in semen?
Fructose is secreted by seminal vesicles. A block in seminal vessels in indicated by the absence of fructose in semen.

11-3. Neonatal hypoglycemia is seen in which conditions?
Glycogen storage disease, type I; galactosemia; fructose intolerance (Box 11.1).

11-4. Congenital cataract is seen in which condition?
Galactosemia.

11-5. Galactosemia is due the absence of which enzyme?
Galactose-1-phosphate uridyl transferase.

11-6. What is the treatment for galactosemia?
Lactose free diet is given for first 5 years of life.

11-7. Why 5 years? why not life-long?
By 5 years, the alternate pathway (galatose-1-phosphate pyrophosphorylase) becomes active.

11-8. Why intake of alcohol produce lactic acidosis?
During alcohol oxidation, NADH is generated, which converts pyruvate to lactate.

11-9. Why excess intake of alcohol produce hypoglycemia?
Because ethanol inhibits gluconeogenesis.

11-10. What are the features of chronic alcoholism?
Polyneuropathy; fatty liver; cirrhosis, encephalopathy.

11-11. In glycoproteins, carbohydrate residues are attached to which group of the polypeptide chain?
Hydroxyl group of serine or threonine.

11-12. Increased glucosaminoglycans in urine is seen in which condition?
Mucopolysaccharidosis.

CHAPTER 12: METABOLISM OF FATTY ACIDS

12-1. Complete digestion of triacylglycerol (triglyceride) needs what?
Pancreatic lipase; co-lipase; Isomerase and bile salts.

12-2. What are the final end products of digestion of triglyceride?
2-monoacylglyceride, 1-monoacylglyceride, glycerol and two fatty acids (Fig.11.1).

12-3. What is the difference for absorption of short chain fatty acid?
Short and medium chain fatty acids do not need re-esterification. They directly enter into blood vessels (Not to lymph vessels).

12-4. Where will you find short and medium chain fatty acids?
Butter, ghee, coconut oil and mother's milk.

12-5. How bile salts help in absorption of lipids?
By emulsifying the lipids and producing micelles of lipids.

12-6. What are micelles?
The micelles are spherical particles with a hydrophilic exterior and hydrophobic interior core. Monoglycerides, long chain fatty acids, cholesterol, and phospholipids are incorporated into molecular aggregates to form mixed micelles.

12-7. What happens to the fatty acids in the mucosal cell?
Once inside the intestinal mucosal cell, the long chain fatty acids are re-esterified to form triglycerides.

12-8. What is the further fate of this triglyceride?
The triglycerides are incorporated into chylomicrons, which are enters the lymph channels.

12-9. What is enterohepatic circulation of bile salts?
They are reabsorbed from the ileum and returned to the liver to be re-excreted again to gut.

12-10. What is steatorrhea?
When daily excretion of fat in feces is more than 6 g per day, it is called steatorrhea. It is due to defective digestion as in chronic diseases of pancreas. In such cases, unsplit fat is excreted in feces.

12-11. What happens in defective absorption?
If the absorption alone is defective, most of the fat in feces may be split fat, i.e. fatty acids and monoglycerides.

12-12. What is cause for defective absorption of fat?
It may be due to diseases in intestinal mucosa, e.g. celiac disease, sprue, Crohn's disease. Any condition leading to a deficiency of bile salts can also result in malabsorption of fat, e.g. gallstones, tumors of head of pancreas, etc.

12-13. How fatty acids are activated in preparation of oxidation?
Fatty acids are activated to their co-enzyme A (CoA) derivative.

12-14. How much ATP is required for this reaction?
One molecule of ATP is hydrolyzed to AMP and PPi. Thus two high energy bonds are utilized in this reaction.

12-15. What are the co-enzymes needed for fatty acid oxidation?
FAD and NAD (Fig. 11.9).

12-16. What is the function of carnitine?
The long chain fatty acyl CoA cannot pass through the inner mitochondrial membrane. Therefore, a transporter, carnitine is involved in transfer of fatty acids.

12-17. What about medium and small chain fatty acids?
Medium chain fatty acids do not require carnitine for transport; so they are easily oxidized.

12-18. What is the net generation of ATP, when one molecule of palmitic acid (16 carbon) is oxidized completely?
106.

12-19. What is the product of beta oxidation of odd chain fatty acids?
Propionyl CoA.

12-20. What is the metabolism of propionyl CoA?
Propionyl CoA is first carboxylated to methyl malonyl CoA then to form succinyl CoA. The succinyl CoA then enters TCA cycle.

12-21. What are the co-enzymes required for the conversion of propionyl CoA to succinyl CoA?
Biotin, ATP, vitamin B_{12}.

12-22. Succinyl CoA is generated from which substances?
Odd chain fatty acids, propionic acid, valine, isoleucine, threonine (Fig. 12.11).

12-23. Succinyl CoA is utilized for what purposes?
Porphyrin biosynthesis; activation of acetoacetate; oxidation in TCA cycle (Fig. 12.11).

12-24. Where is alpha oxidation taking place?
In endoplasmic reticulum (microsomes).

12-25. Refsum's disease is due to what?
Accumulation of phytanic acid, due to defective alpha oxidation.

12-26. What is the rate limiting enzyme of de novo synthesis of fatty acid?
Acetyl CoA carboxylase.

12-27. What is the reaction?
Acetyl CoA + CO_2 to produce Malonyl CoA (Fig. 11.15).

12-28. What are co-enzymes required for the reaction?
Biotin and ATP

12-29. How NADPH is made available?
In the HMP shunt pathway, glucose-6-phosphate dehydrogenase reaction produces NADPH.

12-30. Acetyl CoA is used for what purposes?
Fatty acid synthesis; oxidation in citric acid cycle for generation of energy; cholesterol synthesis and ketone body formation.

12-31. How is fatty acid synthesis regulated?
Key enzyme, acetyl CoA carboxylase, is stimulated by citrate and inhibited by palmitoyl CoA.

12-32. Which hormone enhances the synthesis of triacylglycerol ?
Insulin.

12-33. In adipose tissue, what is the source of glycerol phosphate for triglyceride formation?
From dihydroxyacetone phosphate, derived from glucose.

12-34. White adipose tissue is concerned with what?
Energy storage.

12-35. Brown adipose tissue is involved in what ?
Thermogenesis.

12-36. What enzyme is involved in lipolysis?
By hormone sensitive lipase.

12-37. What is its action?
It hydrolyzes triglyceride into fatty acid.

12-38. In blood, fatty acids are transported as what?
Albumin is the carrier of free fatty acid.

12-39. Which hormones activate hormone sensitive lipase? or, what are the lipolytic hormones?
Growth hormone; corticosteroids; ACTH; adrenalin; glucagon (Fig.12.16).

12-40. In diabetes mellitus, non-esterified fatty acid level in blood is increased; why?
Insulin inhibits hormone sensitive lipase; in diabetes, this inhibition is removed, so, more lipolysis is taking place.

12-41. What are the sources of acetyl CoA?
Pyruvate, fatty acids, acetoacetyl CoA, leucine.

12-42. What substances will prevent fatty liver?
Choline; methionine; lecithin.

12-43. What are ketone bodies?
Acetoacetate; beta hydroxybutyric acid; and acetone.

12-44. Ketone bodies are formed in which tissue?
Liver.

12-45. Ketogenesis is taking place in which subcellular organelle?
Mitochondria.

12-46. What is the rate limiting step in ketone body formation?
HMGCoA synthase.

12-47. Ketone body utilization is taking place in which organs?
Ketolysis is taking place in extra hepatic tissues. (All other tissues, except liver).

12-48. Utilization of ketone bodies by peripheral tissue needs which enzyme?
Succinyl CoA dependent thiophorase.

12-49. What test is used to identify ketone bodies?
Rothera's test.

CHAPTER 13: CHOLESTEROL AND LIPOPROTEINS

13-1. What is thc ring structure present in cholesterol?
Perhydrocyclopentano phenanthrene ring.

13-2. What are substances derived from cholesterol?
Glucocorticoids, mineralocorticoids, testosterone, estrogen, bile acids.

13-3. What is the rate limiting step in the cholesterol biosynthesis?
HMG CoA reductase

13-4. How cholesterol is excreted?
Through bile, partly as cholesterol itself, and partly as bile salts.

13-5. What is the normal level of total cholesterol?
140–200 mg/dL (Table 13.2).

13-6. Maximum cholesterol content is in which lipoprotein?
LDL (Low density lipoprotein) (Table 13.3).

13-7. What is the function of chylomicron?
Transport of triglycerides from intestine to adipose tissue.

13-8. What is the main apoprotein present in chylomicron?
Apoprotein, B48.

13-9. Highest content of triglycerides is seen in which lipoprotein?
Chylomicrons.

13-10. Which carries endogenous triglycerides?
VLDL (Very low densinty lipoprotein).

13-11. What is the main apoprotein present in LDL?
B100; it is the ligand for LDL receptor.

13-12. How LDL is taken up by the cells?
When the apo B-100 binds to the receptor, the receptor-LDL complex is internalized by endocytosis.

13-13. What is the function of LDL?
Transport of cholesterol from liver to peripheral tissues.

13-14. What is "bad cholesterol"?
LDL cholesterol (Fig. 13.14).

13-15. Why is it called so?
LDL transports cholesterol from liver to peripheral tissues, where it is deposited, and causes atherosclerosis.

13-16. What is lipoprotein(a) or Lp(a)?
It has homology with plasminogen. So it interferes with plasminogen activation and impairs fibrinolysis (Fig. 13.15). This leads to unopposed intravascular thrombosis and possible myocardial infarction.

13-17. What is the significance of Lp(a)?
Lp(a) is associated with heart attacks at the age of 30 or 40 years.

13-18. What is the function of HDL?
Transport of cholesterol from peripheral tissues to liver.

13-19. What is "good cholesterol"?
HDL cholesterol.

13-20. Why is it called so?
HDL transports cholesterol from peripheral tissues to liver, and so helps in excretion of cholesterol from the body. So HDL is anti-atherogenic.

13-21. What is the main apoprotein present in HDL?
Apo A-1; it is the ligand for HDL receptor.

13-22. What is LCAT?
Lecithin cholesterol acyl transferase.

13-23. Where is it present?
LCAT present in plasma is activated by apo-A1, when LCAT binds to HDL disc.

13-24. What is its importance?
The free cholesterol is esterified by LCAT; the esterified cholesterol is then incorporated into

HDL. So for excretion of cholesterol, LCAT is necessary.

13-25. What is the importance of PUFA in cholesterol metabolism?
PUFA present in lecithin is transferred to cholesterol by the enzyme LCAT. The esterified cholesterol is then taken by HDL, and finally excreted through liver. So, for excretion of cholesterol PUFA is required. Thus, PUFA will lower the blood level of cholesterol.

13-26. Free fatty acids of plasma are bound to what?
Bound to serum albumin.

13-27. Triglycerides present in adipose tissue are hydrolyzed by what enzyme?
Hormone sensitive lipase.

13-28. Bile acids are derived from what substance?
From cholesterol.

CHAPTER 14: PUFA, PROSTAGLANDINS

14-1. What is MCFA (Medium chain fatty acids)?
They contain 8 to 14 carbon atoms.

14-2. What is LCFA (Long chain fatty acids)?
They contain 16 to 18 carbon atoms.

14-3. What is VLCFA (Very long chain fatty acids)?
They contain 20 or more carbon atoms.

14-4. What is the difference in digestion of MCT from that of LCFA?
MCT containing triglycerides are digested by MCT-specific lipase. Pancreatic lipase and bile salts are not required.

14-5. What is the difference in absorption of MCFA from that of LCFA?
MCFA is absorbed directly to blood (not to lacteals, as in the case of LCFA). After absorption, MCFA is carried by albumin in blood, whereas LCFA are absorbed as triglycerides and carried by chylomicrons.

14-6. When unsaturated fatty acids are oxidized, how many ATP is formed?
The energy yield is less by 1.5 ATP molecules per double bond, when compared to the corresponding chain length saturated fatty acid. Because, the

FAD dependent dehydrogenation (step 1 of beta oxidation) does not occur at the double bond.

14-7. Name polyunsaturated fatty acids.
Linoleic, linolenic and arachidonic acids.

14-8. What are essential fatty acids?
Those cannot be synthesized by the body. So they are to be provided in the diet.

14-9. Name the essential fatty acids.
Linoleic and linolenic acids are the only fatty acids which cannot be synthesized in the body.

14-10. Where is desaturation of fatty acid takes place?
In the endoplasmic reticulum.

14-11. What are the substances derived from PUFA?
Prostaglandins, prostacyclin, thromboxanes, Leukotrienes.

14-12. What are the functions of PUFA?
Synthesis of prostaglandins, synthesis of phospholipids and esterification of cholesterol.

14-13. Prostaglandins are derived from what ?
Prostaglandins are derived from the PUFA. The Series 2 (with 2 double bonds) are derived from Arachidonic acid. All naturally occurring PGs belong to the 2 series.

14-14. What are the enzymes necessary for prostaglandin synthesis?
The arachidonic acid is released by the action of phospholipase A2. Prostaglandin synthesis is catalyzed by prostaglandin H synthase (PGHS). It contains two separate enzyme activities, cyclo-oxygenase and peroxidase.

14-15. How prostaglandin synthesis is regulated?
The phospholipase is activated by epinephrine. Steroids inhibit PL and prevent release of arachidonic acid from membranes.

14-16. What is the importance of cyclo-oxygenase?
Cyclo-oxygenase is activated by catecholamines and inhibited by nonsteroid anti-inflammatory drugs (NSAIDs). Cyclo-oxygenase is a "suicide" enzyme.

14-17. What is the mechanism of action of aspirin?
Aspirin acetylates serine in the active site and irreversibly inhibits the cyclo-oxygenase (Box 14.3).

14-18. What is the mechanism of action of prostaglandins?
Prostaglandins are local hormones, and function through G-protein coupled receptors.

14-19. What is the importance of DHA?
DHA is synthesized linolenic acid. DHA can be obtained from fish oils or from milk. DHA is especially required for the development of brain and retina. Low levels of DHA in blood is seen in patients with retinitis pigmentosa.

14-20. Niemann-Pick disease is due to dcficiency of.
Sphingomyelinase (Table 14.4).

14-21. What is accumulated in Niemann-Pick disease?
Sphingomyelin (Table 14.4).

CHAPTER 15: AMINO ACIDS, GENERAL; UREA CYCLE

15-1. What is the action of pepsin?
It hydrolyses peptide bonds formed by carboxyl groups of phenylalanine, tyrosine, tryptophan and methionine.

15-2. What are zymogens?
They are pro-enzymes; inactive at the time of secretion, but will be activated in the gastro-intestinal tract.

15-3. What is its biological significance?
Zymogens prevent autodigestion of the cells.

15-4. What is its clinical significance?
Acute pancreatitis results when trypsinogen is activated prematurely.

15-5. What are the enzymes in pancreatic juice?
Pancreatic juice contains the endopeptidases trypsin, chymotrypsin and elastase.

15-6. How trypsinogen is activated?
Removal of a hexapeptide from N-terminal end by enterokinase (enteropeptidase).

15-7. What is the action of trypsin?
It hydrolyses peptide bonds formed by carboxyl groups of arginine and lysine.

15-8. How chymotrypsinogen is activated?
It is activated by trypsin.

15-9. What is an endopeptidase?
It acts on peptide bonds inside the protein molecule, so that the protein becomes successively smaller and smaller units.

15-10. Give some examples of an endopeptidase.
Trypsin, pepsin.

15-11. What are exopeptidases?
They act at one end of the protein molecule, liberating amino acids sequentially, one at a time.

15-12. Give an example.
Carboxypeptidase.

15-13. Carboxypeptidase contains which metal?
It is a metallo-enzyme containing zinc.

15-14. In fasting state, nitrogen is transported from muscle as what form?
In fasting state, the muscle releases alanine and glutamine of which alanine is taken up by liver and glutamine by kidneys.

15-15. Transamination of glutamic acid produces what?
Alpha-ketoglutaric acid (Fig. 15.8).

15-16. What is the co-enzyme for transamination reaction?
Pyridoxal phosphate (Fig.15.8).

15-17. What are the physiological significance of transamination?
It synthesizes non-essential amino acids; it is the first step of amino acid breakdown pathways.

15-18. What is the clinical significance of transamination?
Transaminases are increased in blood in liver and cardiac diseases.

15-19. What is transdeamination?
Transamination takes place in all the cells of the body; the amino group is transported to liver as glutamic acid which is finally oxidatively deaminated in liver. Thus the two components of the reaction are physically far away, but physiologically they are coupled. Hence the term transdeamination (Fig. 15.9).

15-20. Which amino acid is oxidatively deaminated in liver?
Glutamic acid (Fig. 15.9).

15-21. Nitrogen atoms in the urea is derived from what precursors?
One from ammonia and another from aspartic acid.

15-22. Ammonia is trapped in brain by what?
Glutamine synthetase (Fig. 15.11).

15-23. What is the key enzyme of urea synthesis?
Carbamoyl phosphate synthetase.

15-24. What are the two carbamoyl phosphate synthetases?
CPS-I is involved in urea synthesis; CPS-II is required for pyrimidine synthesis. CPS-I is seen in mitochondria, while CPS-II is in cytosol (Table 15.2).

15-25. What is the normal blood urea level?
20–40 mg/dL.

15-26. Blood urea level is markedly increased in which condition?
Renal diseases.

15-27. Tetrahydrofolic acid is used for what purpose?
It is the carrier of one carbon compounds.

15-28. Name one carbon compounds?
Formyl, formimino, methenyl, hydroxymethyl, methylene, and methyl (Table 15.4).

15-29. Which are the donors to one carbon pool?
Serine, choline, glycine, tryptophan, histidine (Fig. 15.16).

15-30. One carbon units are used for what?
C2 of purine, C8 of purine, serine, dTMP, choline, creatine, epinephrine (Fig. 15.18).

CHAPTER 16: SIMPLE AMINO ACIDS

16-1. Glycine is used for synthesis of what compounds? Name biologically important substances derived from glycine.
Serine, creatine, purines, heme, glutathione, bile salts.

16-2. For creatine synthesis, which amino acids are used?
Glycine, arginine, methionine.

16-3. What is the significance of creatinine?
Excretion is increased in muscle dystrophy.

16-4. How creatinine is produced in the body?
By spontaneous degradation of creatine phosphate.

16-5. What are the sources of oxalic acid in urine?
Ascorbic acid and glycine.

16-6. What is the cause of hyper oxaluria?
Protein targeting defect.

16-7. Choline is derived from which amino acid?
Serine (Fig.16.12).

16-8. Alanine is transaminated to what?
Pyruvic acid.

16-9. What is significance of glucose-alanine cycle?
During starvation, alanine is released from muscle, and is taken up by liver. In liver alanine is transaminated to pyruvate, and pyruvate undergoes gluconeogenesis.

16-10. Give examples of transmethylation reactions?
Guanido acetic acid to creatine, serine to choline, norepinephrine to epinephrine, N-acetyl serotonin to melatonine (Table 16.1).

16-11. What is the methyl donor in transmethylation reaction? (What is active methionine?)
S-adenosyl methionine.

16-12. What is glutathione ?
Gamma glutamyl cysteinyl glycine.

16-13. What are the functions of glutathione?
Keeping RBC membrane integrity; carrying amino acids across membranes; detoxification of peroxidases.

16-14. What is the test for homocystinuria?
Cyanide nitroprusside test will be positive in urine.

16-15. What is defective enzyme in homocystinuria?
Cystathionine synthase.

16-16. What are the characteristic features of homocystinuria?
Mental retardation, subluxation of lens, thrombosis.

16-17. What is defective enzyme in Cystathionuria?
Cystathionase.

CHAPTER 17: ACIDIC AND BASIC AMINO ACIDS

17-1. What are the major functions of glutamic acid?
Transamination; transdeamination; ammonia trapping in brain; production of gamma aminobutyric acid.

17-2. Which amino acid is oxidatively deaminated in liver?
Glutamic acid.

17-3. Ammonia is trapped in brain by what?
Glutamine synthetase (Fig.17.1).

17-4. What is the reaction catalyzed by glutaminase?
Glutamine to glutamic acid + ammonia (Fig.17.1).

17-5. This reaction is used for what purpose?
Excretion of ammonia in kidney tubules.

17-6. Transamination of glutamic acid will give rise to what?
Alpha ketoglutarate

17.7. Decarboxylation of glutamic acid will give rise to what?
Gamma aminobutyric acid (Fig. 17.2).

17.8. What is GABA?
It is an inhibitory neurotransmitter. It is produced by decarboxylation of glutamic acid.

17.9. Which amino acid will give rise to an inhibitory neurotransmitter?
Glutamic acid gives rise to GABA.

17.10. Glutamic acid can be formed from which amino acids?
Histidine, arginine, proline, glutamine (Fig.17.4).

17.11. Pyruvic acid is transaminated to what?
Alanine.

17.12 Transamination of oxaloacetic acid will give rise to what?
Aspartic acid.

17.13. What are the functions of glutamine?
Ammonia trapping in brain; ammonia production in kidney; synthesis of purine (nitrogen 3 and 9); synthesis of pyrimidine; synthesis of guanine and cytosine (Fig. 17.4).

17-14. Which amino acids are required for both purine and pyrimidine synthesis?
Aspartic acid and glutamine (Fig.17.4).

17-15. What is asparaginae?
Enzyme catalysing the reaction asparaginase to aspartic acid + ammonia (Fig. 17.3).

17-16. What is its clinical application?
It is an anticancer drug against leukemias and lymphomas. The leukemia cells cannot synthesis glutamine, and the available glutamine is destroyed by glutaminase, leukemia cells will starve and die.

17-17. Lysine is deficient in which food stuffs?
Lysine is deficient in cereals.

17-18. Carnitine is synthesized from which amino acid?
Lysine.

17-19. What are the major functions of nitric oxide?
Vasodilatation; keeping the normal blood pressure; inhibitioin of adhesion of platelets, bactericidal in macrophages.

17-20. What is the precursor of nitric oxide?
Arginine.

17-21. Branched chain keto acids are excreted in urine in what condition?
Maple syrup urine disease.

17-22. What is the defect in maple syrup urine disease?
Deficient decarboxylation of branched chain keto acids.

CHAPTER 18: AROMATIC AMINO ACIDS

18-1. What are the important substances synthesized from tyrosine?
Melanin; thyroxine; epinephrine and norepinephrine.

18-2. What is the immediate precursor of norepinephrine?
Dopamine (Fig. 18.4).

18-3. In pheochromocytoma, urine contains what substance?
Vanillyl mandelic acid.

18-4. Phenylalanine level in blood is increased in what condition?
Phenylketonuria.

18-5. What is the defective enzyme in phenylketonuria?
Phenylalanine hydroxylase.

18-6. What are the salient features of phenylketonuria?
Mental retardation, hyperactivity, high blood phenylalanine.

18-7. What is the treatment of phenylketonuria?
Reduce phenylalanine in the diet.

18-8. Homogentisic acid is excreted in urine in which condition?
Alkaptonuria.

18-9. What is the defective enzyme in alkaptonuria?
Homogentisic acid oxidase.

18-10. In phenylketonuria, urine will be positive for what test?
Ferric chloride test.

18-1. Which enzyme is absent in albinism?
Tyrosinase is absent.

18-12. What is difference between tyrosinase and tyrosine hydroxylase?
Both converts tyrosine to DOPA. But tyrosinase is involved in melanin synsthesis, while tyrosine hydroxylase is in catecholamines (Box 18.1).

18-13. What are the important substances produced from tryptophan?
Alanine, acetoacetyl CoA, formyl group, niacin, NAD, serotonin, melatonin.

18-14. What are the excretory products produced from tryptophan?
Hydroxyindole acetic acid and indican.

18-15. Which amino acid will give rise to a vitamin?
Tryptophan gives rise to pyridoxal.

18-16. Serotonin is derived from which amino acid?
Tryptophan.

18-17. What is serotonin?
It is derived from tryptophan. It is a vasoconstrictor.

18-18. What is MAO (Mono amino oxidase)?
It deaminates monoamines. MAO inhibitors are used to treat hypertension and depression.

18-19. Tryptophan is deficient in which food stuff?
Maize and corn.

18-20. Pellagra is manifested in which conditions?
Niacin deficiency; pyridoxal deficiency; tryptophan deficiency; carcinoid syndrome.

18-21. What is Hartnup disease?
Absorption of aromatic amino acids from intestine as well as reabsorption from renal tubules are defective. So tryptophan deficiency, and pellagra like symptoms are seen.

18-22. Tryptophan is excreted in large quantities in which condition?
Hartnup disease.

18-23. What is physiological importance of histidine?
Histidine has a pK value of 6.8; and so it has the maximum buffering capacity at physiological pH? .

18-24. What is the precursor of histamine?
Histidine.

18-25. What is decarboxylation product of histidine?
Histamine.

18-26. What is the clinical significance of histamine?
It is a powerful vasodilator and mediator of anaphylaxis.

18-27. What is FIGLU excretion test?
In folic acid deficiency, there is a block in histidine metabolism, and FIGLU is excreted in large quantities in urine.

18-29. What are partially ketogenic and partially glucogenic amino acids?
Phenylalanine, tyrosine, tryptophan, isoleucine (Fig. 17.14).

18-30. What are important aminoacidurias which cause mental retardation?
Phenylketonuria, homocystinuria, maple syrup urine disease (Table 18.1).

CHAPTER 19: CITRIC ACID CYCLE

19-1. What are the steps in which carbon dioxide is liberated, during oxidation of glucose?
Pyruvate dehydrogenase; isocitrate dehydrogenase; alpha-ketoglutarate dehydrogenase (Fig. 19.2).

19-2. Acetyl CoA is produced from what substrates?
Pyruvate; fatty acids; ketogenic amino acids (Fig.19.1).

19-3. Acetyl CoA is used for what purposes?
Oxidation in TCA cycle; fatty acid synthesis; cholesterol synthesis; ketone body formation.

19-4. Give examples of substrate level phosphorylation.
1,3-bisphosphoglycerate kinase; pyruvate kinase; succinate thiokinase.

19-5. Which is the substrate level phosphorylation step in the TCA cycle?
Succinate thiokinase (Fig. 19.2).

19-6. How many ATPs are generated per one rotation of the citric acid cycle?
10 ATP (Fig. 19.5).

19-7. What is the net yield of ATP from one molecule of glucose in anaerobic glycolysis?
2 ATP (Table 9.4).

19-8. What is the net yield of ATP from one molecule of glucose in aerobic glycolysis?
7 ATP (Table 9.5).

19-9. During complete oxidation, what is the net yield of ATP from one glucose molecule?
32 ATP (Table 9.6).

19-10. What is anaplerosis?
The reactions which involve replenishment of TCA cycle intermediates. For example, pyruvate gives rise to oxaloacetate.

CHAPTER 20: BIOLOGICAL OXIDATION

20-1. Give examples of high energy compounds.
ATP, GTP, creatine phosphate, 1,3-bisphosphoglycerate, phosphoenol pyruvate, acetyl CoA, succinyl CoA (Table 20.2).

20-2. What are the activities taking place inside mitochondria?
Citric acid cycle; electron transport chain; beta oxidation fatty acid.

20-3. Where are enzymes of citric acid cycle located?
Fluid matrix.

20-4. Cytochrome oxidase is present in which complex?
Complex IV.

20-5. What are NAD+ linked dehydrogenases?
Glyceraldehyde-3-phosphate dehydrogenase; pyruvate dehydrogenase; alpha-ketoglutarate dehydrogenase; Isocitrate dehydrogenase; malate dehydrogenase; beta hydroxy acyl CoA dehydrogenase (Fig. 20.13).

20-6. What are FAD linked dehydrogenases?
Succinate dehydrogenase; acyl CoA dehydrogenase.

20-7. Which cytochromes contain copper?
Cytochrome oxidase.

20-8. How many ATP are produced in the oxidation of 1 molecule of NADH?
2.5.

20-9. How many ATP are produced in the oxidation of 1 molecule of FADH?
1.5.

20-10. Which is complex V of respiratory chain?
ATP synthase.

20-11. What is valinomycin?
It acts as an ionophore; dissipates the proton gradient; and so inhibits ATP synthesis.

20-12. What is atractylocide?
It inhibits translocase; and inhibits ATP synthesis.

20-13. What is oligomycin?
It inhibits oxidative phosphorylation.

20.14. Name inhibitors of oxidative phosphorylation.
Oligomycin; carbon monoxide; cyanide.

20-15. What is the cause for death due to cyanide poisoning?
Cyanide inhibits cytochrome oxidase.

CHAPTER 21: HEME

21-1. What is the rate limiting enzyme of heme synthesis?
ALA synthase.

21-2. Methenyl bridge of protoporphyrin is derived from what?
Alpha carbon of glycine.

21-3. How is heme synthesis regulated?
ALA synthase is repressed by heme.

21-4. What is the action of barbiturates on heme synthesis?
Barbiturates will induce heme synthesis.

21-5. What is the defect in acute intermittent porphyria?
PBG-deaminase.

21-6. How porphyrins are detected?
They show fluorescence under ultraviolet light.

21-7. In acute intermittent porphyria, urine contains.
ALA and PBG.

21-8. Degradation of heme needs which enzyme?
Heme oxygenase system, with NADPH.

21-9. Degradation of heme releases what gas?
Carbon monoxide.

21-10. Heme is converted to bilirubin in which site?
Microsomes of reticuloendothelial cells.

21-11. Bilirubin in blood is carried by what?
Albumin.

21-12. Where is the conjugation taking place?
In liver.

21-13. What is the enzyme needed for conjugation?
UDP-glucuronyl transferase.

21-14. What is the normal level of plasma bilirubin?
0.2–0.8 mg/dL.

21-15. What is latent jaundice?
When plasma bilirubin is between 1 to 2 mg/dL.

21-16. What is jaundice?
When plasma bilirubin is more than 2 mg/dL, it diffuses into tissues, producing yellowish discoloration of tissues.

21-17. Enterohepatic circulation is seen in which substances?
Urobilinogen and bile salts.

21-18. What is the defect in Crigler-Najjar syndrome?
Defect in conjugation of bilirubin due to deficiency of UDP glucuronyl transferase.

21-19. Bilirubin in serum is estimated by what test?
van den Bergh reaction.

21-20. What is direct van den Bergh reaction?
The color is developed immediately when blood is added.

21-21. What is the inference, when direct van den Bergh test is positive?
Blood contains conjugated bilirubin; it is water soluble.

21-22. What is indirect van den Bergh test?
When blood is added to the solution, there is no color; but when alcohol is added, color is developed.

21-23. What is kernicterus?
In young children, when plasma bilirubin is more than 20 mg/dL, it diffuses into brain, causing permanent damage to brain cells.

21-24. In obstructive jaundice, blood contains what?
Conjugated bilirubin in excess quantity.

21-25. Bile salts and bile pigments are excreted in urine in which condition?
Obstructive jaundice.

CHAPTER 22: HEMOGLOBIN

22-1. Hemoglobin molecule is made up of what?
Hb has 4 sub units, 2 alpha and 2 beta units. It contains 4 iron atoms.

22-2. How many molecules of oxygen can bind with hemoglobin?
Hb can bind 4 molecules of oxygen.

22-3. 100 mL of blood can carry how much oxygen?
20 mL.

22-4. What is Bohr effect?
The influence of pH and pCO_2 to facilitate oxygenation of Hb in the lungs and deoxygenation at the tissues is known as the Bohr effect.

22-5. What is chloride shift?
When CO_2 is taken up, chloride ions from the plasma would enter. This is called chloride shift or Hamburger effect.

22-6. How hemoglobin carries CO_2?
75% as isohydric transport and 15% as carbamino-hemoglobin.

22-7. What are the forms in which carbon dioxide is transported?
Dissolved form; isohydric transport; and carbamino hemoglobin.

22-8. What is isohydric transport of carbon dioxide?
There is minimum change in pH during the transport.

22-9. How is this effected?
The H^+ ions are buffered by the deoxy-Hb.

22-10. What is the function of 2,3 BPG?
It will decrease the affinity of hemoglobin for oxygen.

22-11. What is the difference between HbA and HbF?
HbA has 2 alpha and 2 beta chains, but HbF has 2 alpha and 2 gamma chains.

22-12. What is the physiological significance of HbF?
Oxygen affinity is more for HbF than HbA. HbF is seen in fetal circulation.

22-13. What is the clinical significance of HbF?
It is seen in adults in hemoglobinopathies and thalassemias.

22-14. Why carbon monoxide becomes a poison?
Hb has more affinity to carbon monoxide than oxygen.

22-15. What is met-hemoglobin?
Hemoglobin in which iron is in ferric state or in oxidized form.

22-16. What is hemoglobin S (HbS)?
The glutamic acid in the 6th position of beta chain of HbA is changed to valine in HbS.

22-17. What is the cause for sickle cell anemia?
Solubility of deoxy HbS is lower than deoxy HbA; so HbS is precipitated intracellularly, leading to sickle shape of RBC.

22-18. How HbS is identified?
HbS is slower moving on electrophoresis than HbA.

22-19. What is sickle cell trait?
Heterozygous (AS) condition; One allele is normal; the other is abnormal; so half of Hb molecules are normal; and half abnormal.

22-20. What is its clinical significance?
Persons with sickle cell trait will not have any disease manifestations. But, at higher altitudes, hypoxia may cause manifestation of the disease.

CHAPTER 23: CLINICAL ENZYMOLOGY, BIOMARKERS

23-1. Lactate dehydrogenase has how many iso-enzymes?
Five. They are H4, H3M, H2M2, M3H and M4 varieties.

23-2. What is meant by flipped pattern?
Normally LDH-2 (H3M1) concentration in blood is greater than LDH-1 (H4); but this pattern is reversed in myocardial infarction; this is called flipped pattern.

23-3. What serum markers are helpful in the diagnosis of myocardial infarction?
Creatine kinase (CK) CK MB isoenzyme and troponins.

23-4. When does creatine kinase (CK) level increased?
Myocardial infarction; muscular dystrophies.

23-5. What is the advantage of CK estimation?
CK level starts to rise within 3 hours of infarction. Therefore CK estimation is very useful to detect early cases.

23-6. What are the isoenzymes of CK?
CK is a dimer; the subunits are called B for brain and M for muscle. Therefore three iso-enzymes are possible.

23-7. What is the advantage of cardiac troponin I?
Cardiac troponin I is released into the blood within 4 hours after the onset of cardiac symptoms, peaks at 12–16 hours. Therefore, CTI is very useful as a marker at any time interval after the heart attack.

23-8. Alkaline phosphatase level in serum is elevated in which conditions?
Moderate increase is seen in hepatic diseases (infective hepatitis); high levels in obstructive

jaundice; and very high levels in bone diseases such as Paget's disease, rickets, metastatic carcinoma of bone.

23-9. What is Regan isoenzyme?
It is the isoenzyme of alkaline phosphatase. It is of placental origin. It is elevated in about 15% cases of carcinoma of lung, liver and gut. Regan isoenzyme is otherwise known as carcinoplacental isoenzyme.

23-10. Estimation of gamma glutamyl transferase is useful to detect which condition?
Alcohol abuse.

23-11. What are the enzymes useful in diagnosing liver pathology?
ALT, ALP, GGT, NTP.

23-12. Serum acid phosphatase level is increased in which condition?
Prostate carcinoma.

23-13. What is the use of isoenzyme study of acid phosphatase?
Tartarate labile isoenzyme is specific for prostate carcinoma.

23-14. What is advantage of prostate specific antigen?
PSA is very specific for prostate carcinoma.

23-15. What are enzymes useful as tumor markers?
(a) Regan isoenzyme of ALP for lung tumor; (b) Tartarate labile iso-enzyme of ACP; (c) Prostate specific antigen (PSA) for prostate carcinoma; and (d) Neuron specific enolase (NSE) for cancers of neuroendocrine origin.

23-16. Which enzyme deficiency is X-linked?
Glucose-6-phosphate-dehydrogenase.

23-17. How the deficiency of GPD is manifested?
Drug-induced hemolytic anemia.

23-18. Acute pancreatitis can be diagnosed by estimating which enzymes?
Amylase and lipase.

CHAPTER 24: REGULATION OF BLOOD GLUCOSE

24-1. What is the normal level of fasting blood sugar?
70–110 mg/dL.

24-2. What are the sources of blood glucose?
(a) Absorption from intestine; (b) Hepatic gluconeogenesis; and (c) Hepatic glycogenolysis (Fig. 24.3).

24-3. Which hormone is hypoglycemic?
Insulin.

24-4. What are the major actions of insulin?
Insulin decreases blood sugar; it stimulates glycolysis; inhibits gluconeogenesis; enhances glycogen synthesis and inhibits lipolysis.

24-5. What are the hyperglycemic hormones?
Glucagon, adrenaline, corticosteroids, growth hormone.

24-6. What are the major actions of glucagon?
Promotes glycogenolysis; enhances gluconeogenesis, depresses glycogen synthesis, inhibits glycolysis.

24-7. What is the major indication for doing an oral glucose tolerance test (OGTT)?
Patient has symptoms suggestive of diabetes mellitus; but fasting blood sugar value is inconclusive (between 110 and 126 mg/dL).

24-8. What are criteria to diagnose diabetes mellitus?
1. Fasting plasma sugar is more than 126 mg/dL, on more than one occasion. 2. Or, if 2-hr post-glucose load value of OGTT is more than 200 mg/dL (even at one occasion). 3. Or, if both fasting and 2-hour values are above these levels, on the same occasion. 4. If HbA1c is more than 5.5%

24-9. What is impaired glucose tolerance (IGT)?
When fasting plasma glucose level is between 110 and 126 mg/dL or 2-hour post-glucose value is between 140 and 200 (Table 24.1).

24-10. What is gestational diabetes mellitus (GDM)?
This term is used when carbohydrate intolerance is noticed, for the first time, during a pregnancy.

24-11. What is the clinical significance of GDM?
Women with GDM are at increased risk for subsequent development of frank diabetes. GDM is associated with increased birth weight of child and increased incidence of neonatal mortality.

24-12. What is renal glucosuria?
Glucose is excreted in urine due to a lowering of renal threshold. The blood sugar levels are within normal limits.

24-13. What is normal renal threshold for glucose?
180 mg/100 mL.

24-14. What are reducing substances seen in urine?
Glucose, fructose, lactose, galactose, pentoses, ascorbic acid, glucuronides.

24-15. Fructosuria is due to what?
Due to the deficiency of fructokinase or aldolase B.

24-16. What is lactosuria?
It is observed in the urine of normal women during 3rd trimester of pregnancy and during lactation.

24-17. What is the clinical importance of lactosuria?
The condition is harmless. But it is important to distinguish lactosuria from glucosuria.

24-18. What is the test for reducing sugars in urine?
Benedict's test.

24-19. Where is insulin synthesized?
Beta cells of Langerhans of pancreas.

24-20. What is pro-insulin?
Insulin is synthesized as a large single polypeptide. Middle part of it is then removed, to form the A and B chains of insulin (Fig. 24.5).

24-21. What are the pathways stimulated by insulin?
Glycolysis, glycogen synthesis, HMP shunt pathway, lipogenesis (Table 24.4).

24-22. What are the enzymes stimulated by insulin?
Phosphofructokinase; glycogen synthase, glucose-6-phosphate dehydrogenase; acetyl CoA carboxylase.

24-23. What are the pathways inhibited by insulin?
Gluconeogenesis, glycogenolysis, lipolysis, ketogenesis.

24-24. What are the enzymes inhibited by insulin?
Glucose-6-phosphatase,glycogen phosphorylase, hormone sensitive lipase.

24-25. How is insulin secretion controlled?
Glucose is the major stimulant of insulin secretion.

24-26. How glucose stimulates insulin secretion?
GluT 2 receptors act as sensor mechanism for glucose level. Cyclic AMP along with calcium causes the insulin secretion.

24-27. Which tissue utilizes glucose maximally?
At basal rates, brain utilizes 60% of sugar oxidized.

24-28. What are the cardinal symptoms of diabetes mellitus?
Polyuria, polydypsia, polyphagia, weight loss.

24-29. What is the reason for polyuria in diabetes ?
When the blood glucose level exceeds, the renal threshold glucose is excreted in urine. Due to osmotic effect, more water accompanies the glucose.

24-30. What is the reason for polydypsia in diabetes?
To compensate for this loss of water, thirst centre is activated, and more water is taken (polydypsia).

24-31. What is the reason for weight loss in diabetes?
The loss and ineffective utilization of glucose leads to breakdown of fat and protein. This would lead to loss of weight.

24-32. What is the reason for polyphagia in diabetes?
To compensate the loss of glucose and protein, patient takes more food.

24-33. What is micro-albuminuria?
Albumin 30 to 300 mg/day in urine. It is a predictor of progressive renal damage, atherosclerotic diseases and cardiovascular mortality.

24-34. What is the difference between glycosylation and glycation?
Enzymatic addition of any sugar to a protein is called "glycosylation", while non-enzymatic process is termed "glycation".

24-35. What is significance of glycated hemoglobin?
An elevated glycohemoglobin indicates poor control of diabetes mellitus.

CHAPTER 25: CARDIOVASCULAR DISEASES

25-1. What are the salient features of hyperlipoproteinemia Type II-A?
Premature atherosclerosis; elevated plasma LDL cholesterol; prominent beta band on electrophoresis.

25-2. What is it due to?
Defect in LDL receptor.

25-3. Hypercholesterolemia is seen in what conditions?
Diabetes mellitus; nephrotic syndrome; Obstructive jaundice; hypothyroidism.

25-4. What are the important risk factors of coronary artery diseases?
Serum cholesterol level above 200 mg/dL; LDL-cholesterol level above 160 mg/dL; HDL-cholesterol level below 35 mg/dL; Apo(a) above 30 mg/dL.

25-5. What are other risk factors associated with coronary artery diseases?
Cigarette smoking, hypertension, diabetes mellitus, serum triglyceride level above 200 mg/dL; homocysteine level; sedentary lifestyle, obesity.

25-6. What advise you will give to a person with increased cholesterol level?
Reduce cholesterol content of food; include PUFA and omega-3 fatty acids in diet; reduction of total fat intake; increase green leafy vegetables; exercise.

CHAPTER 26: LIVER FUNCTION TESTS

26-1. What is the normal serum bilirubin level?
0.2 to 0.8 mg/dL.

26-2. What is normal level of unconjugated bilirubin?
0.2–0.6 mg/dL.

26-3. What is normal level of conjugated bilirubin?
Less than 0.2 mg/dL.

26-4. What is latent jaundice?
Serum bilirubin between 1 mg/dL and 2 mg/dL.

26-5. Jaundice appears at what level of bilirubin?
Jaundice appears if the serum bilirubin goes above 2 mg/dL.

26-6. How jaundice is classified?
(i) Pre-hepatic or hemolytic. (ii) Hepatocellular. (iii) Post-hepatic or obstructive jaundice (Table 26.4).

26-7. What type of bilirubin is present in hemolytic jaundice?
Unconjugated bilirubin is increased in serum. Hence van den Bergh test is indirect positive (Table 26.4).

26-8. What happens in obstructive jaundice?
In obstructive jaundice, conjugated bilirubin in serum is elevated, and van den Bergh test is direct positive.

26-9. What is the picture in hepatocellular jaundice?
In hepatocellular jaundice, a biphasic reaction is observed, because both conjugated and unconjugated bilirubins are increased in serum (Table 26.4).

26-10. What are important liver function tests?
Serum bilirubin, albumin, alkaline phosphatase; urine bile salts, bile pigments and urobilinogen (Tables 26.2 and 26.4).

26-11. What are the features of obstructive jaundice?
Elevated conjugated bilirubin in plasma; presence of bile salts and bile pigments in urine; increased alkaline phosphatase level in blood; increased cholesterol in blood.

26-12. Which tests will be positive in a urine of a patient with obstructive jaundice?
Gmelin's test, Hay's test, Fouchet's test (Table 26.4).

26-13. What are the enzymes useful in diagnosis of liver diseases?
Alanine aminotransferase; alkaline phosphatase; gamma glutamyl transferase (Table 26.1).

26-14. What is the characteristic laboratory finding in chronic alcoholism?
Elevation of gamma glutamyl transferase level in serum.

26-15. What is alkaline tide?
When HCl is produced in stomach, bicarbonate level within the cell increases (formed from

H_2CO_3), it is reabsorbed into bloodstream. This would account for the alkaline tide of plasma and urine, immediately after meals.

CHAPTER 27: KIDNEY FUNCTION TESTS

27-1. What is normal glomerular filtration rate?
120–125 mL per minute.

27-2. What is the function of glomerul us?
An ultrafiltrate of the blod is produced in glomerulus, while the cells and proteins are retained in the blood.

27-3. What is the normal specific gravity of urine?
1.015 to 1.025.

27-4. What is the minimum and maximum specific gravity of urine?
1.003 to 1.032.

27-5. What is the specific gravity of urine in chronic renal failure?
Fixed to 1.010.

27-6. Specific gravity of urine increased when?
Diabetes mellitus and acute glomerulonephritis.

27-7. Polyuria is seen in which conditions?
Diabetes mellitus; diabetes insipidus; chronic renal failure.

27-8. Proteinuria is seen in which conditions?
Acute glomerulonephritis; nephrotic syndrome; pyelonephritis; multiple myeloma.

27-9. What is the test to detect urinary protein?
Heat and acetic acid test.

27-10. What is micro-albuminuria?
When small quantities of albumin (30–300 mg/day) is seen in urine.

27-11. What are non-protein nitrogen (NPN)?
Urea, uric acid, creatinine. In kidney diseases, the levels of these substances in blood are increased.

27-12. What is meant by renal clearance?
Clearance is defined as the quantity of blood or plasma completely cleared of a substance per unit time and is expressed as milliliter per minute. It is the milliliter of plasma which contains the amount of that substance excreted by the kidney within a minute.

27-13. What is the best method for assessing glomerular filtration rate?
Creatinine clearance test.

27-14. What is the advantage of creatinine test?
Creatinine is formed spontaneously (non-enzymatic), so the blood level and excretion rate of creatinine is a constant.

27-15. What is the significance of creatinine clearance?
A decreased creatinine clearance is a very sensitive indicator of a reduced glomerular filtration rate.

27-16. What is the normal creatinine level in blood?
For adult males, 0.7–1.4 mg/dL;
for adult females, 0.6–1.3 mg/dL.

27-17. What is creatinine coefficient?
It is the urinary creatinine expressed in mg/kg body weight.

27-18. Urea clearance is lowered in which condition?
Chronic liver failure.

27-19. What is the normal blood urea level?
20–40 mg/dL.

27-20. What is the clinical significance of urea level?
Urea level is increased in renal failure.

27-21. How is tubular concentrating capacity measured?
By measuring urine specific gravity.

27-22. What are the tests to assess tubular function?
Measurement of specific gravity, concentration test, dilution test, acidification test.

CHAPTER 28: PLASMA PROTEINS

28-1. What is normal serum albumin level?
3.5 to 5 mg/dL.

28-2. What is normal value of total proteins in serum?
6–8 g/100 mL.

28-3. Edema due to hypoproteinemia is seen in which conditions?
Cirrhosis liver; malnutrition; nephrotic syndrome.

28-4. Albumin carries what substances?
Free fatty acids; bilirubin; salicylate; calcium.

28-5. What are the functions of albumin?
It maintains colloidal osmotic pressure of plasma; it transports non-esterified fatty acid and bilirubin.

28-6. Albumin is synthesized in which organ?
Liver.

28-7. Where are gamma globulins synthesized?
By reticuloendothelial system (spleen, lymph nodes).

28-8. Albumin-globulin ratio is reversed in which conditions?
Cirrhosis; chronic infections; nephrotic syndrome; multiple myeloma.

28-9. Name some transport proteins.
Retinol binding protein, thyroxine binding globulin, transcortin (cortisol); haptoglobin (hemoglobin); transferrin (iron); hemopexin (free heme) (Table 28.1).

28-10. Name some acute phase proteins.
C-reactive protein; ceruloplasmin; haptoglobin.

28-11. Polymorphism is exhibited by which proteins?
Haptoglobin; transferrin; ceruloplasmin.

28-12. What is the clinical manifestation of Alpha-1-anti-trypsin deficiency?
Emphysema and chronic lung infections.

28-13. What is ceruloplasmin?
It is a copper containing enzyme (ferroxidase) seen in blood. It is an acute phase protein.

28-14. What is the significance of ceruloplasmin?
Ceruloplasmin level in blood is decreased in Wilson's hepatolenticular degeneration.

28-15. What is the carrier protein of copper?
Albumin.

28-16. Hemopexin carries what?
Free heme (Table 28.1).

28-17. How hemophilia is transmitted?
It is inherited as an X-linked recessive trait; males are affected; females are carriers.

CHAPTER 29: ACID-BASE BALANCE

29-1. When pH falls by 1 unit, what is the change in the hydrogen ion concentration?
Increases 10 times.

29-2. Relationship between pH and pK is given by which equation?
Henderson-Hasselbalch's equation.

29-3. What determines the pH of buffer?
By the ratio of salt to acid.

29-4. What determines the buffering capacity?
The absolute concentrations of the salt.

29-5. What is meant by buffering capacity?
The buffering capacity of a buffer is defined as the ability of the buffer to resist changes in pH when an acid or base is added.

29-6. Buffer is most effective at which pH?
When the pH is nearer to the pK of the acid.

29-7. In the blood, which buffer is most effective?
Bicarbonate buffer.

29-8. What are the mechanisms for maintaining the normal pH of plasma?
Buffers of plasma; lung mechanism; kidney mechanism.

29-9. What is the alkali reserve of the body?
Bicarbonate is the alkali reserve.

29-10. What is the normal ratio of bicarbonate to carbonic acid in blood?
Bicarbonate to carbonic acid ratio is 20.

29-11. What is the respiratory mechanism of pH regulation?
In acidosis, there is hyperventillation, so CO_2 is lowered, and H_2CO_3 is lowered.

29-12. Glutaminase enzyme is used for what purpose?
For production of ammonia in kidney tubules (Fig.29.5).

29-13. What is metabolic acidosis?
Primary deficit of bicarbonate.

29-14. What are the causes of metabolic acidosis?
Diabetic ketosis, chronic renal failure, diarrhea.

29-15. What are the features of diabetic ketoacidosis?
Lowered bicarbonate; elevated plasma chloride; increased anion gap.

29-16. How to calculate anion gap?
(sodium + potassium) minus (chloride + bicarbonate).

29-17. What is the cause for high anion gap acidosis?
Diabetic ketoacidosis, chronic renal failure, renal tubular acidosis, lactic acidosis.

29-18. What is metabolic alkalosis?
Primary excess of bicarbonate.

29-19. What are the causes of metabolic alkalosis?
Prolonged vomiting; gastric aspiration; and ingestion of antacids.

29 20. What is respiratory acidosis?
Primary excess of carbonic acid.

29-21. What are the causes of respiratory acidosis?
Bronchial asthma; bronchopneumonia; narcotic poisoning.

29-22. What is respiratory alkalosis?
Primary deficit of carbonic acid.

29-23. What is the cause for respiratory alkalosis?
Hyperventillation, as in hysteria, salicylate poisoning.

CHAPTER 30: ELECTROLYTE AND WATER BALANCE

30-1. What is the major intracellular cation?
Potassium (Table 30.2).

30.2. What is the major extracellular cation?
Sodium (Table 30.2).

30-3. What are the factors regulating fluid and electrolyte balance?
Aldosterone, ADH, renin angiotensin, thirst.

30-4. What is major cause for isotonic contraction?
Small intestinal obstruction.

30-5. What are the causes for hypotonic contraction?
Infusion of dextrose (without saline), Addison's disease.

30-6. What are the causes for hypertonic contraction?
Diarrhea, vomiting.

30-7. What are the causes for isotonic expansion of extracellular fluid?
Congestive cardiac failure and hyperaldosteronism.

30-8. What is the cause for hypotonic expansion of extracellular fluid?
Inappropriate secretion of ADH.

30-9. What is the important cause for hypertonic expansion of ECF?
Cushing's syndrome.

30-10. What are the metabolic imbalances seen in diarrhea?
Metabolic acidosis; hypertonic contraction of ECF; urine with high specific gravity; urine output reduced.

CHAPTER 31: BODY FLUIDS

31-1. How lactose is synthesized?
The galactose unit is transferred from UDP-galactose to glucose. This is catalyzed by lactose synthase.

31-2. What is the major difference between human and cow's milk?
Human milk has higher carbohydrate content than cow's milk while protein content is less (Table 31.1).

31-3. Milk contains which type of fatty acids?
The fatty acids are mainly saturated, but 50% of them are medium chain fatty acids (lauric and myristic acids).

31-4. What is the advantage of MC fatty acids?
They are easily digested, absorbed and metabolized.

31-5. What is the major protein in cow's milk?
80% protein of cow's milk is casein.

31-6. What type of protein is casein?
It is a phosphoprotein.

31-7. How the phosphate group is attached to protein?
The phosphate groups are added to the hydroxyl groups of serine or threonine residues.

CHAPTER 33: FREE RADICALS AND ANTIOXIDANTS

33-1. What are the enzymes generating ROS in macrophages?
NADPH oxidase, superoxide dismutase, myeloperoxidase.

33-2. What are the free radical scaviging enzymes?
Superoxide dismutase; glutathione peroxidase, glutathione reductase; and catalase.

33-3. Name some diseases related with ROS?
Chronic inflammation, rheumatoid arthritis, bronchopulmonary dysplasia, respiratory distress syndrome, retrolental fibroplasia, cataract, reperfusion injury, atherosclerosis, peptic ulcer.

33-4. Name important anti-oxidants.
Alpha tocopherol (vitamin E), ascorbic acid (vitamin C), vitamin A, beta carotene.

CHAPTER 35: SEPARATION TECHNIQUES

35-1. What is electrophoresis?
The term refers to the movement of charged particles through an electrolyte when subjected to an electric field.

35-2. Electrophoresis is commonly employed for what purpose in laboratory?
For serum electrophoresis. To see abnormalities in serum protein concentrations.

35-3. What is the principle of adsorption chromatography?
Separation is based on differences in adsorption at the surface of a solid stationary medium.

35-4. What is partition chromatography?
The components of the mixture to be separated are partitioned between the two phases depending on the partition co-efficient (solubility) of the particular substances.

35-5. What are the common types of partition chromatography?
Paper chromatography and thin layer chromatography.

35-6. What is the basic principle of ion-exchange chromatography?
The separation is based on electrostatic attraction between charged molecules to oppositely charged groups on the ion exchange resins (Fig. 54.10).

35-7. What is the quickest method for separation of proteins?
HPLC (high performance liquid chromatography).

35-8. What is the advantage of radioimmunoassay?
Very small quantities of substances could be accurately measured.

35-9. What are the disadvantages of RIA, when compared to ELISA?
1. Since radioisotopes are used, only approved laboratories could take up the assay. 2. The shelf life of the reagent is short.

35-10. What are the enzymes commonly used in ELISA technique?
Horse radish peroxidase (HRP) and alkaline phosphatase (ALP).

CHAPTER 36: FAT SOLUBLE VITAMINS

36-1. What is the major function of vitamin A?
Important in vision, used in the Wald's visual cycle.

36-2. What causes the nerve impulse in retina?
Photo-isomerisation of 11-cis retinal to all trans retinal.

36-3. How is alltrans-retinal regenerated?
Trans-retinal is taken to liver, where it is made to trans-retinol, then isomerised to cis-retinol and then to cis-retinal.

36-4. What enzymes are required for regeneration?
Alcohol dehydrogenase and retinol isomerase. (Fig. 36.4).

36-5. What are the deficiency manifestations of vitamin A?
Night blindness, xerophthalmia, keratomalacia, keratinization of epithelium.

36-6. What are the sources of vitamin A?
Carrot, mangoes, papaya, green leafy vegetables, fish oil.

36-7. What is the daily requirement of vitamin A?
750 to 1000 microgram.

36-8. How is cholecalciferol synthesized?
From 7-dehydrocholesterol in the Malpighian layer of epidermis, by the action of ultraviolet rays.

36-9. How is vitamin D activated?
Cholecalciferol from skin reaches liver. There it is hydroxylated to form 25-hydroxycholecalciferol

(25-HCC). It then reaches kidney, where further hydroxylation takes place to form 1, 25-dihydroxy-cholecalciferol (DHCC).

36-10. What is calcitriol?
1,25-dihydroxycholecalciferol, or active vitamin D, contains three hydroxyl groups at 1, 3 and 25 positions (Fig. 36.9).

36-11. Which vitamin acts as a pro-hormone?
Vitamin D is converted to calcitriol.

36-12. What is the function of vitamin D?
It increases absorption of calcium from intestine; it also increases mineralization of bone.

36-13. What are deficiency manifestions of vitamin D?
Rickets in children and osteomalacia in adults.

36-14. In renal disease, oral doses of vitamin D may not be effective, why?
Hydroxylation and activation of vitamin is taking place in kidney.

36-15. What is the daily requirement of vitamin D?
10–15 microgram.

36-16. What is the function of vitamin E?
It is the most powerful anti-oxidant.

36-17. How selenium and vitamin E are related?
They act synergistically as anti-oxidants.

36-18. What is the source of vitamin E?
Vegetable oils, e.g. wheat germ oil, sunflower oil, safflower oil, cotton seed oil, palm oil.

36-19. What is the daily requirement of vitamin E?
15 mg or 33 international units.

36-20. What is the function of vitamin K?
Gamma carboxylation of clotting factors such as prothrombin

36-21. Deficiency of vitamin K can occur in which conditions?
Obstructive jaundice; antibiotic therapy; administration of dicoumarol.

36-22. What is the mechanism of action of dicoumarol?
It competitively inhibits vitamin K epoxide reductase.

36-23. Dicoumarol is used for what purpose?
To prevent intravascular thrombosis.

36-24. Excess dose of vitamin K in neonates may lead to which condition?
Hemolysis and jaundice.

36-25. Bleeding tendency is common in deficiency of.
Vitamin K; vitamin C; platelets; prothrombin.

CHAPTER 37: WATER SOLUBLE VITAMINS

37-1. What is the source of thiamine?
Aleurone layer of cereals (food grains) is a rich source of thiamine. Whole wheat flour and unpolished rice and yeast are other good sources.

37-2. Thiamine pyrophosphate is required for?
Transketolase; pyruvate dehydrogenase; alpha-ketoglutarate dehydrogenase.

37-3. Which vitamin is required for oxidative decarboxylation?
Thiamine pyrophosphate.

37-4. In thiamine deficiency, what alterations are seen in the blood?
Increased pyruvic acid level and increased transketolase.

37-5. What are the clinical manifestations of thiamine deficiency?
Beriberi, Wernick syndrome, polyneuritis.

37-6. Beriberi is due to deficiency of which vitamin?
Thiamine.

37-7. What is the daily requirement of thiamine?
1 to 1.5 milligram.

37-8. What is the co-enzyme function of riboflavin?
FMN and FAD.

37-9. Name some FAD dependent enzymes.
Succinate dehydrogenase; acyl CoA dehydrogenase; xanthine oxidase; glutathione reductase; pyruvate dehydrogenase; alpha-ketoglutarate dehydrogenase.

37-10. What are the manifestations of riboflavin deficiency?
Glossitis, cheilosis, angular stomatitis, circumcorneal vascularization.

37-11. What are the dietary sources of riboflavin?
Rich sources are liver, dried yeast, egg and milk.

37-12. What is the daily requirement of riboflavin?
1.5 mg per day.

37-13. Which vitamin is synthesized in the body?
Niacin.

37-14. What are the co-enzyme function of niacin?
NAD+ and NADP+

37-15. Name some NAD+ dependent enzymes.
Lactate dehydrogenase; glyceraldehyde-3-phosphate dehydrogenase; Pyruvate dehydrogenase; beta hydroxyacyl CoA dehydrogenase; mitochondrial isocitrate dehydrogenase.

37-16. Name the NADPH generating reactions.
Glucose-6-phosphate dehydrogenase; 6-phosphogluconate dehydrogenase.

37-17. What are important NADPH utilizing reactions?
Beta ketoacyl ACP dehydrogenase; alpha, beta unsaturated ACP dehydrogenase; HMGCoA reductase; met-hemoglobin reductase; dihydrofolate reductase (Box 37.5).

37-18. Pellagra is seen in deficiency of which vitamin?
Niacin.

37-19. What are the salient features of pellagra?
Dermatitis, diarrhea, dementia.

37-20. What is the precursor of niacin?
Tryptophan.

37-21. Tryptophan will give rise to how much niacin?
About 60 mg of tryptophan will yield 1 mg of niacin.

37-22. Tryptophan is deficient in which food stuff?
Maize and corn.

37-23. What are the conditions that will lead to symptoms of pellagra?
Isoniazid therapy; Low tryptophan content in diet; niacin deficiency; hartnup disease; carcinoid syndrome.

37-24. What is the dietary sources of niacin?
Rice polishing, cereals, legumes, meat and fish contain niacin. About half of the requirement is met by the conversion of tryptophan to niacin.

37-25. What is the daily requirement of niacin?
The RDA is 20 mg/day.

37-26. Transamination reaction requires which vitamin?
Pyridoxal phosphate.

37-27. Pyridoxal phosphate is required for what reactions?
Transamination, decarboxylation of amino acids, ALA synthase, glycogen phosphorylase.

37-28. What is clinical significance of transaminase?
Blood level of ALT is increased in liver diseases.

37-29. Give examples of decarboxylation reactions.
Glutamate to GABA (gamma aminobutyric acid); histidine to histamine; 5-hydroxytryptophan to serotonin; serine to ethanolamine.

37-30. What are the manifestations of pyridoxal deficiency?
Infantile convulsions; peripheral neuritis; pellagra; anemia.

37-31. What is the reason for infantile convulsions in pyridoxal deficiency?
Pyridoxal phosphate is required for the formation of GABA; it is an inhibitory neurotransmitter, absence of which leads to hyper-excitation and convulsions.

37-32. What is the reason for peripheral neuritis in pyridoxal deficiency?
PLP is involved in the synthesis of sphingolipids; so B_6 deficiency leads to demyelination of nerves and consequent peripheral neuritis.

37-33. What is the reason for pellagra like disease in pyridoxal deficiency?
Niacin is produced from tryptophan. One enzyme of this pathway is PLP dependent. So B_6 deficiency in turn leads to niacin deficiency. It is manifested as pellagra.

37-34. Give an exmple of one vitamin deficiency leading to another vitamin deficiency?
PLP deficiency in turn leads to niacin deficiency which is manifested as pellagra.

37-35. What is the reason for anemia in pyridoxal deficiency?

PLP is required for ALA synthase. So, hypochromic microcytic anemia may occur due to the inhibition of heme biosynthesis.

37-36. What are the drugs which cause pyridoxal deficiency?

Isonicotinic acid hydrazide (INH) (isoniazid); cycloserine, and oral contraceptives act as B_6 antagonists.

37-37. What are the dietary sources of pyridoxal?

Rice polishing, wheat germs, cereals, legumes (pulses), egg, milk, meat, fish and green leafy vegetables.

37-38. What is the daily requirement of pyridoxal phosphate?

1 to 2 mg/day.

37-39. What is co-enzyme form of pantothenic acid?

Co-enzyme A.

37-40. What is the function of CoA?

Oxidation of pyruvic acid and activation of acyl groups.

37-41. Deficiency of pantothenic acid leads to what?

Burning foot syndrome.

37-42. What are the donors for acetyl CoA pool?

Pyruvate, fatty acid, amino acids.

37-43. Acetyl CoA is used for what purposes?

Oxidation in TCA cycle; fatty acid synthesis; cholesterol synthesis, steroid synthesis, ketone body formation.

37-44. Succinyl CoA is used for what purposes?

Oxidation in TCA cycle; heme synthesis, activation of acetoacetate (Fig. 37.13).

37-45. What is the function of biotin?

Carboxylation reactions.

37-46. Name biotin dependent carboxylation reactions.

Acetyl CoA carboxylase; propionyl CoA carboxylase; pyruvate carboxylase.

37-47. Name carboxylation reactions which are not dependent on biotin.

Carbamoyl phosphate synthetase, malic enzyme.

37-48. What is the antagonist for biotin?

Avidin.

CHAPTER 38: WATER SOLUBLE VITAMINS CONTINUED

38-1. What is the co-enzyme form of folic acid?
Tetrahydrofolic acid.

38-2. What is the main function of folic acid?
Tetrahydrofolic acid is the carrier of one carbon units.

38-3. What are the causes of folate deficiency?
Pregnancy; defective absorption; anticonvulsant drugs (dilantin, phenobarbitone); hemolytic anemias, dietary deficiency.

38-4. What is the major manifestation of folic acid deficiency?
Macrocytic anemia.

38-5. What is the complication of folic acid deficiency in pregnancy?
Folic acid deficiency during pregnancy may lead to neural tube defects (such as spina bifida) in the fetus.

38-6. What is the relatioin of folic acid deficiency and cardiac diseases?
Folic acid is useful to reduce the level of homocysteine in blood, and thus helps in preventing heart diseases.

38-7. Laboratory findings in folic acid deficiency?
(1) Normal folic acid level in serum is decreased. (2) FIGLU excretion test positive. (3) Peripheral blood picture shows macrocytic anemia. (4) Homocysteine level in blood is increased.

38-8. What is the daily requirement of folic acid?
The RDA of folate is 200 mg/day. In pregnancy the requirement is increased to 400 mg/day.

38-9. What is the mechanism of action of sulphonamides?
They have structural similarity with PABA. Therefore they competitively inhibit the enzyme responsible for the incorporation of PABA into folic acid. So, they kill bacteria.

38-10. What is mechanism of action of methotrexate?
It inhibits folate reductase, and is a powerful anticancer drug.

38-11. What are inhibitors of folic acid?
Methotrexate, pyrimethamine, sulphonamide.

38-12. Name a water soluble vitamin, which is stored in the body.
Vitamin B_{12}.

38-13. What is the metal present in vitamin B_{12}?
Cobalt.

38-14. Vitamin B_{12} is absorbed from where?
Ileum.

38-15. How vitamin B_{12} is absorbed?
With the help of Intrinsic factor of Castle.

38-16. What is the carrier of vitamin B_{12} in blood?
Transcobalamin-II, is the specific carrier.

38-17. What is co-enzyme function of vitamin B_{12}?
Methyl malonyl CoA isomerase; homocysteine methyl transferase.

38-18. What are the causes of vitamin B_{12} deficiency?
Nutritional gastrectomy, pernicious anemia, atrophy of gastric epithelium.

38-19. What is pernicious anemia?
Auto-antibodies against intrinsic factor inhihits vitamin B_{12} absorption from intestine.

38-20. A patient who has undergone gastrectomy is likely to develop deficiency of which vitamin?
Vitamin B_{12}.

38-21. What are the important manifestations of vitamin B_{12} deficiency?
Megaloblastic anemia and subacute combined degeneration.

38-22. What abnormal compound is excreted in vitamin B_{12} deficiency?
Methyl malonic acid in urine.

38-23. What is folate trap?
When B_{12} is deficient, methyl THFA to THFA reaction cannot take place. So folate is trapped.

38-24. What is the clinical importance of folate trap?
There is associated folic acid scarcity in B_{12} deficiency.

38-25. What is the clinical difference in folate deficiency and B_{12} deficiency?
In both conditions, there is macrocytic anemia, in B_{12} deficiency, there is additional neurological symptoms.

38-26. What is the daily requirement of vitamin B_{12}?
1 to 2 microgram per day.

38-27. In vitamin B_{12} deficiency, what are the abnormalties seen in urine?
Urine may contain Methyl malonic acid; homocystine; cystathione; formimino glutamic acid.

38-28. Which is the vitamin which is totally absent in plant sources?
Vitamin B_{12}.

38-29. What are the sources of vitamin B_{12}?
Liver, meat, fish, curd.

38-30. What is the chemical structure of vitamin C?
L-ascorbic acid.

38-31. Is vitamin C synthesized in the body?
Man and primates cannot synthesize ascorbic acids. Lower animals could synthesize it from glucose through glucuronic acid pathway.

38-32. What is the major function of ascorbic acid?
It promotes collagen formation through its action on post-translational hydroxylation of proline and lysine residues.

38-33. Which vitamin is required for post-translational modifications?
Ascorbic acid.

38-34. What is the daily requirement of ascorbic acid?
75 mg per day.

38-35. What is the disease manifested in vitamin C deficiency?
Scurvy.

38-36. What are the important features of scurvy?
Hemorrhagic tendency, microcytic anemia, bone pain, bleeding gums.

CHAPTER 39: MINERAL METABOLISM

39-1. What influences absorption of calcium from intestine?
Vitamin D; calcitriol; parathyroid hormone; phytic acid; oxalate.

39-2. What is the daily requirement of calcium?
500 mg per day.

39-3. What are the sources of calcium?
Milk, egg, fish, vegetables.

39-4. What is the function of calcium ?
Coagulation; neuromuscular activity; intracellular messenger; activation of enzymes; bone formation.

39-5. What is the normal level of calcium in blood?
9–11 mg/dL.

39-6. Which will influence serum calcium level?
Calcitriol; calcitonin; parathyroid hormone.

39-7. How parathyroid hormone regulates calcium level in blood?
Bone resorption; increased absorption of calcium from intestines; increased absorption of calcium from renal tubules.

39-8. Hypocalcemia results in what condition?
Tetany.

39-9. What is the cause for tetany?
Hypoparathyroidism.

39-10. What are the features of hyperparathyroidism?
Osteoporosis; hypercalciuria; urinary calculi.

39-11. What is the normal level of phosphorus in blood?
3–4 mg/dL.

39-12. What is the normal level of sodium in blood?
136–145 mEq/L.

39-13. What is the major extracellular cation?
Sodium.

39-14. Which hormones control the sodium level?
ADH, aldosterone, cortisone.

39-15. What is the normal blood level of potassium?
3.5 to 5 mEq/L.

39-16. What is the major intracellular cation?
Potassium.

39-17. What is the manifestation of hypokalemia?
In ECG, T wave inverted, ST segment is lowered.

39-18. What is the normal level of chloride in blood?
96–106 mEq/L.

39-19. What is the daily requirement of iron for a normal adult male?
10–20 milligram per day.

39-20. What are the dietary sources of iron?
Green leafy vegetables; jaggery.

39-21. Which trace element is deficient in milk?
Iron.

39-22. What are important iron containing proteins?
Hemoglobin, myoglobin, cytochromes, catalase, tryptophan pyrrolase, xanthine oxidase, transferrin, ferritin.

39-23. Which will increase iron absorption from intestines?
Gastric HCl, ascrobic acid, cysteine.

39-24. What are the factors which will retard iron absorption?
Phytic acid in cereals; oxalic acid in leafy vegetables; calcium, zinc, lead and phosphates.

39-25. How is iron absorbed?
Iron in the ferrous state is binding to divalent metal transporter (DMT-1) present in mucosal surface of intestinal cell; thus iron is transported into the mucosal cell (Fig. 39.9).

39-26. What is mucosal block?
Iron homeostasis is maintained by regulation at the level of absorption and not by excretion. When iron stores in the body are depleted, absorption is enhanced. When adequate quantity of iron is stored, absorption is decreased. This is referred to as "mucosal block".

39-27. How is this mucosal regulation takes place?
Absorption of iron needs DMT and ferroportin. Synthesis of both these proteins is downregulated by hepcidin.

39-28. What is hepcidin?
Hepcidin is secreted by liver, when body iron reserves are adequate. In anemia, the synthesis of hepcidin is reduced.

39-29. What is the carrier protein in iron in blood?
Transferrin.

39-30. What is the storage form of iron?
Ferritin.

39-31. What is hemosiderin?
Excess iron is loaded as hemosiderin.

39-32. What is haptoglobin?
It is the carrier of free hemoglobin.

39-33. What is hemopexin?
It is the carrier of free heme.

39-33. Anemia is resulted in the deficiency of.
Iron, copper, vitamin C, folic acid, vitamin B_{12}, pyridoxal.

39-34. What are the causes of iron deficiency?
Nutritional, hookworm infection, repeated pregnancy, chronic blood loss, nephrosis.

39-35. How iron deficiency manifests?
Microcytic hypochromic anemia.

39-36. What are the features of hemosiderosis?
Cirrhosis of liver; diabetes mellitus; yellow color of skin.

39-37. What is the major cause for hemosiderosis?
Repeated transfusion of whole blood.

39-38. Name some copper containing enzymes.
Ceruloplasmin, cytochrome oxidase, cytochrome C, tyrosinase, lysyl oxidase, super oxide dismutase.

39-39. What are the dietary sources of copper?
Cereals, meat, liver.

39-40. What is ceruloplasmin?
It is ferroxidase; it promotes oxidation of ferrous ion to ferric form.

39-41. What are the characteristic features of Wilson's hepatolenticular degeneration?
Ceruloplasmin level in blood is decreased; copper excretion is reduced; copper is accumulated in liver to produce cirrhosis.

39-42. What is the daily requirement of iodine?
150–200 microgram.

39-43. What is the daily requirement of zinc?
15–20 milligram.

39-44. Which enzymes contain zinc?
Alkaline phosphatase; amylase; carbonic anhydrase; RNA polymerase. Zinc is required for insulin secretion.

39-45. What is the importance of selenium?
It is an anti-oxidant.

39-46. Name the selenium containing enzyme.
Glutathione peroxidase.

CHAPTER 40: NUTRITION

40-1. What is the calorific value of carbohydrates?
4 kilocalories per gram.

40-2. How much calories are generated from fat?
9 kcal per gram of fat.

40-3. What is the respiratory quotient?
It is the ratio of carbon dioxide produced to the oxygen consumed.

40-4. What is the respiratory quotient of carbohydrates?
One.

40-5. The respiratory quotient for a mixed diet.
About 0.82.

40-6. What is specific dynamic action (SDA)?
Increased heat production after intake of food. It is the extra heat produced by the body, over and above the calorie value of the given food. It is also called thermogenic action of food.

40-7. Which foodstuff has maximum specific dynamic action?
Proteins.

40-8. BMR (Basal metabolic rate) is increased in which conditions?
Fever; starvation; cold climate; increased thyroid hormones.

40-9. Increased basal metabolic rate is observed in which clinical condition?
Grave's disease (hyperthyroidism).

40-10. What are the beneficial effects of dietary fiber?
Increased motility of intestine; prevents constipation; decreased absorption of cholesterol; increased glucose tolerance.

40-11. Which food stuffs contain polyunsaturated fatty acids (PUFA)?
Vegetable oils, such as sunflower oil, groundnut oil.

40-12. Name polyunsaturated fatty acids.
Linoleic acid; linolenic acid; arachidonic acid.

40-13. What is the dietary advice for a patient with hypercholesterolemia?
They should reduce intake of food containing cholesterol and increase intake of polyunsaturated fatty acids.

40-14. What foodstuffs contain high cholesterol?
Egg, butter, animal fat, meat (Table 40.8).

40-15. What is the recommended daily allowance of protein for a normal adult?
1 g/kg body weight

40-16. Negative nitrogen balance is observed in which conditions?
Chronic infection; old age, malnutrition.

40-17. Positive nitrogen balance is observed in which conditions?
Pregnancy; convalescence, growth period.

40-18. Lysine is deficient in which foodstuff?
Pulses.

40-19. Phenylalanine is deficient in which food stuff?
Tapioca.

40-20. Methionine is deficient in which food stuff?
Cereals.

40-21. How all amino acids are made available in a mixed diet?
Mutual supplementation by combining cereals and pulses.

40-22. What is protein calorie malnutrition (PCM)?
Deficiency of protein causes kwashiorkor. Deficiency of calorie produces marasmus. Both are two spectra of PCM.

40-23. What are the salient features of kwashiorkor?
Hypoalbuminemia, hypomagnesemia, growth retardation, lethargy, loss of appetite.

40-24. How much energy is required for a 60 kg person, with sedentary work?
2000 kcal.

40-25. A balanced diet should have calories for carbohydrate, proteins and fats in which ratio?
60 : 20 : 20.

40-26. What is the optimum ratio for cereals and pulses in a balanced diet?
Cereals: pulses ratio = 5:1.

40-27. What is the recommended ratio of saturated, monounsaturated and polyunsaturated fatty acids in a balanced diet?
SFA : MUFA : PUFA = 1 : 1 : 1 ratio.

40-28. What is the recommended fat intake?
The ideal fat intake is about 20% of total calories, out of which about 30% be of PUFA.

CHAPTER 41: DETOXIFICATION

41-1. What is detoxification process?
Biochemical processes whereby noxious substances are rendered less harmful or more water soluble and easily excretable.

41-2. What are xenobiotics?
They are compounds accidentally ingested or taken as drugs or compounds produced in the body by bacterial metabolism.

41-3. What are the processes by which body detoxify obnoxious compounds?
Oxidation, reduction, hydrolysis, conjugation.

41-4. Give an example of detoxification by oxidation.
Toluene and alcohol.

41-5. Give an example of detoxification by reduction.
Para-nitrophenol.

41-6. Give examples of detoxification by hydrolysis.
Aspirin to salicylic acid + acetate.

41-7. Give names of conjugating agents.
UDP glucuronic acid; glutathione; glycine, PAPS.

41-8. Give examples of substances detoxified by conjugation.
Bilirubin to bilirubin glucuronide; phenol to phenyl sulphate; benzoic acid to hippuric acid.

41-9. Which amino acids are used for detoxification?
Glutamine; glycine; cysteine.

41-10. What is the sulphating agent?
PAPS (Phosphoadenosyl phosphosulphate).

41-11. How benzoic acid is detoxified?
Benzoic acid + glycine = benzoyl glycine or hippuric acid.

41-12. What is the methylating agent?
S-adenosyl methionine.

CHAPTER 42: HEAVY METAL POISONING

42-1. What is most common environmental poison?
Lead.

42-2. What are common causes of lead poisoning?
Paint, lead containig petrol, lead pipes, newspapers and cigarette smoke are important contaminants.

42-3. What are the occupations in which persons are prone to get lead poisoning?
Battery repair, radiator repair, soldering, painting and printing.

42-4. What are the manifestations of chronic lead poisoning?
In children, mental retardation, learning disabilities, behavioral problems are seen. Anemia and loss of appetite are very common.

42-5. What is the cause of anemia in lead poisoning?
Lead inhibits delta aminolevulinic acid (ALA) synthase, ALA-dehydratase and ferrochelatase. So, heme synthesis is blocked.

42-6. What is Minamata diseases?
Organic mercury poisoning.

CHAPTER 43: NUCLEOTIDES

43-1. What is a nucleotide?
Nitrogenous base + sugar + phosphate.

43-2. What is a nucleoside?
Nitrogenous base + sugar.

43-3. What are the sugars in nucleic acids?
Ribose in RNA and deoxyribose in DNA.

43-4. What are the bases present in nucleotides?
Purines and pyrimidines.

43-5. Name the common purines.
Adenine and guanine.

43-6. Name the common pyrimidines.
Cytosine, uracil, thymine.

43-7. What are the bases present in DNA?
Adenine, guanine, cytosine, thymine.

43-8. Which base is absent in DNA?
Uracil.

43-9. Which base is found exclusively in DNA and not in RNA?
Thymine.

43-10. Which base is found exclusively in RNA?
Uracil.

43-11. Which amino acid is required for both purine and pyrimidine synthesis?
Aspartic acid and glutamine (Figs 43.10 and 43.17).

43-12. N3 of purine ring is donated by what?
Glutamine (Fig. 43.10).

43-13. Glycine donates what part of the purine ring?
C4, C5, N7 atoms.

43-14. What is the key enzyme of de novo synthesis pathway of purines?
Phosphoribosyl amidotransferase (step 1).

43-15. How is de novo synthesis of purine regulated?
Amidotransferase enzyme is inhibited by AMP and GMP.

43-16. Give a few examples of purine analogues, used as anti-cancer drugs.
6-mercapto-purine; cytosine arabinoside; methotrexate; azaserine.

43-17. What is the end product of catabolism of purines in human beings?
Uric acid (Fig. 43.15).

43-18. What is xanthine oxidase?
It is the enzyme catalysing the reactions, hypoxanthine to xanthine and xanthine to uric acid.

43-19. Name an enzyme containing molybdenum.
Xanthine oxidase contains FAD, molybdenum and iron.

43-20. What is the normal uric acid level in blood?
2–5 mg/dL in females and 3–7 mg/dL in males.

43-21. What is gout?
It is a disease associated with hyperuricemia, deposit of urate crystals in joints and excrutiating joint pain.

43-22. Which property of uric acid is responsible for the manifestations of gout?
Uric acid is sparing soluble in water.

43-23. Increased uric acid level is seen when?
Gout, preeclampsia, Hodgkin's lymphoma, leukemia.

43-24. Hyperuricemia is observed in which conditions?
Gout, Lesch Nyhan syndrome, von Gierke's disease.

43-25. What is the mechanism of allopurinol?
It is an analogue of hypoxanthine. It inhibits xanthine oxidase, thereby decreasing the formation of uric acid.

43-26. It is what type of inhibition?
Xanthine oxidase converts allopurinol to alloxanthine. It is a more effective inhibitor of xanthine oxidase. This is a good example of 'suicide inhibition'.

43-27. What is the reason for severe combined immunodeficiency?
Adenosine deaminase deficiency.

43-28. What are starting materials for pyrimidine synthesis?
Carbamoyl phosphate and aspartic acid (Fig. 43.17).

43-29. What is the rate limiting step in pyrimidine synthesis?
Aspartyl transcarbamoylase.

43-30. How is pyrimidine synthesis pathway regulated in mammals?
CPS II is inhibited by CTP.

43-31. Orotic aciduria is a feature of deficiency of which enzymes?
OMP decarboxylase; OPRTase; ornithine transcarbamoylase.

43-32. What are the characteristic features of orotic aciduria?
Megaloblastic anemia.

43-33. What is the mechanism of methotrexate?
It inhibits dihydrofolate reductase and thereby reduces the regeneration of THFA; it is a powerful anticancer agent.

43-34. What is the mechanism of 5-fluorouracil?
It inhibits conversion of dUMP to dTTP; and acts as an antimetabolite.

CHAPTER 44: DNA STRUCTURE, REPLICATION

44-1. What are the salient features of Watson-Crick model of DNA?
(i) Right handed double helix; (ii) Each turn of helix has 10 base pairs; (iii) Hydrogen bonds between purines and pyrimidines; (iv) DNA strands running in opposite directions (anti-parallel).

44-2. What is base pairing rule (Chargaff's rule)?
Pairing (hydrogen bonding) of adenine with thymine and guanine with cytosine. Purine is paired with pyrimidine. A+T = G+C.

44-3. What are histones?
They are nuclear proteins found in nucleus. They bind with DNA and stabilizes DNA structure.

44-4. What is a nucleosome?
DNA wrapped around histones.

44-5. What is meant by DNA replication?
During cell division, each daughter cell gets an exact copy of the genetic information of the mother cell. This process of copying the DNA is known as DNA replication.

44-6. What is semi-conservative replication?
Each one of the newly synthesized DNA has one-half parental strand (old) and one-half new DNA strand. Since half of the original DNA is conserved in daughter DNA, it is called semi-conservative replication.

44-7. What are the enzymes required for DNA replication?
DNA polymerase, topoisomerase, DNA ligase.

44-8. What is a replisome?
DNA replication needs more than 20 enzymes and proteins, collectively called DNA replicase system or replisomes.

44-9. What is a replication bubble?
Helicases move on both directions, separating the strands in advance of the replication. This forms a replication bubble.

44-10. Replication is in which direction?
Polymerisation of the new strand of DNA is taking place from 5´ to 3´ direction. This means that the template is read in the 3´ to 5´ direction.

44-11. How replication starts?
An RNA primer, about 100–200 nucleotides long, is synthesized by the RNA primase.

44-12. What is meant by semi-discontinuous nature of replication?
DNA synthesis is always in the 5′ to 3′ direction in both strands. In one strand, the replication is taking place continuously; but in the other strand replication is in small pieces (Fig. 44.14).

44-13. What is meant by the lagging strand?
The strand which is discontinuously synthesized is referred to as the "lagging strand" and the one continuously polymerized as the "leading strand" (Fig. 44-14).

44-14. What are Okazaki fragments?
The small DNA molecules attached to its own primer RNA in the lagging strand are called Okazaki fragments.

44-15. Xeroderma pigmentosum is due to deficiency of what process?
Defect in DNA repair mechanism (nucleotide excision repair).

44-16. In prokaryotes, DNA replication is inhibited by what drugs?
Ciprofloxacine, nalidix acid, novobiocin.

44-17. DNA replication in eukaryotes is inhibited by what drugs?
5-fluorouracil; 6-mercaptopurine; cytosine arabinoside; etoposide.

44-18. What is telomere?
The end piece of the chromosome is called telomere.

44-19. What is telomerase?
It replicates the end piece of chromosomes.

44-20. Whate is the clinical significance of telomerase?
In old age, the telomerase activity is lost; leading to chromosome instability and cell death. As cancer cells have increased and persistent activity of telomerase, the cancer cells become immortal.

CHAPTER 45: TRANSCRIPTION

45-1. What is transcription?
The process of making a complementary mRNA copy of DNA.

45-2. The mRNA is a complementary copy of which strand of DNA?
Template strand.

45-3. What is the coding strand?
The opposite strand of template strand. Coding strand has the same sequence as of the mRNA (Fig. 45.3).

45-4. What is the enzyme necessary for transcription?
DNA dependent RNA polymerase or RNAP.

45-5. What is the direction of transcription?
It takes place in the 5′ to 3′ direction.

45-6. What is TATA Box?
It is a signal for initiation of transcription in prokaryotes. It is about 10 bp upstream of starting of mRNA synthesis.

45-7. What is the corresponding signal in mammals?
Godberg-Hogness Box.

45-8. What are post-transcriptional modifications?
Removal of introns; addition of a cap at 5′ end; adding poly A tail at 3′ end; and methylation.

45-9. What is intron?
Part of mRNA that is removed is called intron.

45-10. What is a spliceosome?
Small nuclear ribonucleoprotein particles (SnRNPs) associated with hnRNA at the exon-intron junction form spliceosomes (Fig. 1.11).

45-11. What are ribozymes?
Enzymes made up of RNA are called ribozymes.

45-12. Give some examples of ribozymes.
Spliceosomes, RNAse-P (which generates the ends of tRNAs) and peptidyl transferase (present in ribosomes) are examples of ribozymes.

45-13. Name inhibitors of RNA synthesis
Rifampicin, actinomycin, mitomycin and amanitin.

45-14. What is the mechanism of action of mitomycin?
Actinomycin D and mitomycin intercalate with DNA strands, thus blocking transcription. They are used as anticancer drugs.

45-15. What is the mechanism of action of rifampicin?
It binds to beta subunit of RNA polymerase, and inactivates it. Rifampicin is widely used in the treatment of tuberculosis and leprosy.

45-16. What is reverse transcriptase?
It will make a new DNA strand based on the RNA template. Thus genetic information is transferred from RNA to DNA.

CHAPTER 46: TRANSLATION

46-1. What are structural features of tRNA molecule?
It has clover shape appearance. Amino acid binding is at 3´ end. The opposite part has anticodon arm.

46-2. What is a codon?
Codon is a triplet of bases, present in mRNA. Collection of these codons make up the dictionary of genetic code.

46-3. During replication, DNA is synthesized in which direction?
From 5´ to 3´ direction.

46-4. During transcription, mRNA is synthesized in which direction?
From 5´ to 3´ direction.

46-5. During translation, protein is synthesized in which direction?
From amino terminal end to carboxy terminal end.

46-6. What are the salient features of genetic code?
(i) They are consecutive three bases pairs in mRNA. (ii) Codons are non-overlapping; (iii) degenerate; (iv) but unambiguous and (v) universal.

46-7. What is meant by degeneracy of genetic code?
One amino acid is represented by multiple codons.

46-8. What is meant by the term wobbling?
Anticodons pair with codons that differ at the third base.

46-9. Where is protein biosynthesis taking place?
Ribosomal assembly either attached to endoplasmic reticulum, or in cytoplasm.

46-10. Give examples for post-translational modifications.
Gamma carboxylation of prothrombin; hydroxylation of proline in collagen; methylation of histones; glycosylation of proteins.

46-11. Give examples of inhibitors of translation in eukaryotic cells?
Puromycin; cycloheximide; diphtheria toxin; ricin.

46-12. Give examples of reversible inhibitors of protein synthesis in bacteria.
These antibiotics are bacteriostatic. Tetracyclins; chloramphenicol; erythromycin and clindamycin are examples.

46-13. Give an example of irreversible inhibitor of protein synthesis in bacteria.
These antibiotics are bactericidal. Streptomycin is an example.

CHAPTER 47: MOLECULAR GENETICS

47-1. Give examples of X-chromosome linked transmission.
Hemophilia; glucose-6-phsophate dehydrogenase deficiency; duchenne type muscular dystrophy.

47-2. What is a mutation?
An alteration in the genetic material results in a mutation.

47-3. Give an example.
HbS or sickle-cell hemoglobin is produced by a mutation of the beta chain in which the 6th position is changed to valine, instead of the normal glutamate.

47-4. Give examples for mutagens.
Acridine orange; X-rays; gamma rays; methyl cholanthrene.

47-5. What is apoptosis?
Programmed cell death.

47-6. What is cell cycle?
Cell cycle refers to the sequence of events that occur during the growth and cell division.

47-7. In which phase of the cell cycle, DNA synthesis is maximum?
During S phase.

47-8. What is the p53?
It is an oncosuppressor gene product.

47-9. What is an operon?
One or more structural genes, together with their operator gene (Fig. 47.7).

47-10. Give examples of enzyme induction.
Beta galactosidase by lactose; tryptophan pyrrolase by glucocorticoid and ALA synthase by barbiturates.

47-11. Give an example of repression.
ALA synthase by heme.

CHAPTER 48: RECOMBINANT DNA TECHNOLOGY

48-1. What are restriction endonucleases?
They act as "molecular scissors". These enzymes recognise specific sequences in the DNA, and then cleave at those sites. They are useful in recombinant DNA technology.

48-2. What are required for preparing a recombinant DNA molecule?
Restriction endonuclease; plasmid vector; DNA ligase.

48-3. What are plasmids?
They are small circular, extra chromosomal DNA present in bacteria. They are used as vectors in DNA recombinant technology.

48-4. What are the vectors used for gene therapy?
Retrovirus; adenovirus; plasmid, liposome complex.

48-5. Name some diseases in which gene therapy is used successfully?
Severe combined immunodeficiency; Duchenne muscular dystrophy; cystic fibrosis; hemophilia

CHAPTER 49: MOLECULAR BIOLOGY TECHNIQUES

49-1. What is the use of Southern blotting?
To identify abnormal genes, to demonstrate virus integration. Also used for prenatal diagnosis.

49-2. What is Northern blotting?
The Northern blot is used to demonstrate specific RNA.

49-3. What is Western blotting?
This is to identify proteins (not nucleic acids) (Fig. 49.3).

49-4. What is meant by DNA probes?
They are single stranded DNA, labelled with radioactivity. They are used to identify specific bases sequences (genes, virus, etc.) in the cellular DNA material.

49-5. RFLP (restriction fragment length polymorphism) is used for what?
Locating mutations in DNA.

49-6. What is the application o f DNA fingerprinting?
It has medicolegal application.

49-7. Which enzyme is required for PCR (polymerase chain reaction)?
Taq polymerase.

49-8. What is the use of PCR?
Diagnosis of bacterial and viral diseases; medico-legal cases; diagnosis of genetic disorders, especially prenatal diagnosis. Only very minute quantity of sample is required.

49-9. How are monoclonal antibodies produced in the laboratory?
By means of hybridoma technology.

CHAPTER 50: MECHANISM OF ACTION OF HORMONES

50-1. What are G proteins?
They are involved in signal transduction.

50-2. How the G proteins work?
The receptors are in membrane; bin ding of hormone causes activation of G protein (Fig. 50.3).

50-3. How is G protein activated?
Binding of hormone on receptor causes attachment of GTP to G protein, and thereby G protein is activated (Fig. 50.4).

50-4. What is the function of activated G protein?
Activated G protein activates adenyl cyclase (Fig. 50.4).

50-5. What is the function of adenyl cyclase?
Adenyl cyclase will produce cyclic AMP.

50-6. How cyclic AMP is destroyed?
By phosphodiesterase.

50-7. What is the function of protein kinase?
It phosphorylates enzymes or target proteins, so that they are activated.

50-8. Name some hormones which act through cyclic AMP as the second messenger.
Glucagon, ACTH, TSH, ADH, FSH, LH (Table 50.1).

50-9. Name some second messengers.
Cyclic AMP; 1,2-diacyl glycerol; inositol triphosphate; calcium.

50-10. Name the hormones that bind to intracellular receptors.
Glucocorticoids, mineralocorticoids, estrogens, progesterone, androgens, and thyroxin.

CHAPTER 51: PITUITARY HORMONES

51-1. What are the major effects of growth hormone?
GH increases the uptake of amino acids by cells; enhances protein synthesis, produces positive nitrogen balance, lipolysis and hyperglycemia. It is anabolic.

51-2. Increased secretion of growth hormone will lead to which condition?
Excess secretion of GH leads to gigantism in children and acromegaly in adults.

51-3. What is the result of decreased growth hormone secretion?
Pituitary dwarfism.

51-4. What is the function of FSH?
FSH stimulates growth of ovarian follicles in females and spermatogenesis (Sertoli cells) in males.

CHAPTER 52: STEROID HORMONES

52-1. Which is the precursor of steroid hormones?
Cholesterol.

52-2. What are the hormones produced from progesterone?
Corticosterone; aldosterone; testosterone; estrogens.

52.3. Cholesterol has how many carbon atoms?
27

52-4. Cortisol has how many carbon atoms?
21

52-5. Testosterone has how many carbon atoms?
19.

52-6. Estrogen has how many carbon atoms?
18.

52-7. What is the immediate precursor of estrogens?
Testosterone.

CHAPTER 53: THYROID HORMONES

53-1. What is the precursor of thyroxin?
Tyrosine.

53-2. What is the ratio of T4 and T3 in blood?
Blood T4 is 70 times more than T3.

53-3. What is the function of TSH?
It increases uptake of iodine; enhances the oxidation of iodine to iodide; and favors the hydrolysis of thyroglobulin to produce T4.

53-4. What are the functions of thyroid hormones?
Calorigenic effect. BMR is increased. Basal metabolic rate (BMR) is increased by increasing the cellular metabolism.

53-5. Deficiency of thyroxine results in which?
Myxedema and cretinism.

53-6. What are the thyroid function tests?
Assay of T3 and T4; free T3 and T4; plasma TSH; TRH response test; thyroid antibodies.

53-7. What are salient features of hypothyroidism?
Decreased T3 level; increased TSH level; lethargy; hypercholesterolemia; weight gain; decreased BMR.

53-8. What are the characteristic features of hyperthyroidism?
High T4 levels; increased rate of metabolism, weight loss, tachycardia, fine tremors, sweating, diarrhea, emotional disturbances, anxiety and sensitivity to heat.

53-9. What is the cause for Grave's disease?
Auto-antibodies binds to TSH receptor,mimicking continuous stimulation of TSH.

CHAPTER 57: CANCER

57-1. Name anti-mutagens and anti-carcinogens.
Vitamin A, vitamin C, vitamin E, curcumin.

57-2. What are oncogenes?
Genes capable of causing cancer. Oncogenes are specific sequences in DNA which when expressed may produce cancer.

57-3. What are tumor markers?
They are factors released from the tumor cells, which could be detected in blood and therefore indicate the presence of the tumor in the body.

57-4. What is clinical application of tumor markers?
They are useful for: (1) To follow-up cancer and to monitor the effectiveness of the therapy; (2) To detect the recurrence of the tumor.

57-5. Name some commonly used tumor markers.
Alpha fetoprotein; carcinoembryonic antigen; prostate specific antigen (Table 57.8).

57-6. Alpha-fetoprotein (AFP) level is used as a tumor marker for which cancer?
Hepatoma.

57-7. What is the marker for colorectal cancers ?
Carcinoembryonic antigen.

57-8. What is the significance of beta chain of human chorionic gonadotropin?
It is a tumor marker for choriocarcinoma.

57-9. What is the mechanism of action of mitomycin?
Intercalation with DNA strands.

57-10. What is the mechanism of action of methotrexate?
It is a folic acid antagonist.

CHAPTER 58: TISSUE PROTEINS

58-1. What is the structure of collagen?
It has triple stranded quarter staggered arrangement. One-third residues are glycine. About 1% residues are proline and hydroxyproline.

58-2. What are the post-translational modifications taking place in collagen?
Hydroxylation of proline and lysine residues; hydrolysis of pro-collagen.

58-3. Hydroxylation of proline needs what?
It depends on vitamin C.

58-4. In ascorbic acid deficiency, what happens?
Poor hydroxylation; defective collagen and scurvy.

58-5. What is the enzyme for cross linkage formation?
Lysyl oxidase.

58-6. What is its clinical application?
Lysyl oxidase contains copper. So, in copper deficiency lysine cross link formation is deficient, leading to defective collagen.

Part 5: Multiple Choice Questions (MCQs)

Multiple choice questions. Single response type. Each of the questions below is followed by a few suggested answers. Select the one best response among them.

CHAPTER 2: SUBCELLULAR ORGANELLES

2-1. Which of the following cellular organelle serve as the site of protein synthesis?

A. Golgi bodies
B. Endoplasmic reticulum
C. Mitochondria
D. Peroxisomes

2-2. Lysosomes are:

A. Power house of the cell
B. Bags of hydrolytic enzymes
C. Store house for genetic information
D. Interconnected channels in the cytoplasm

2-3. Which of the following cellular organelle can cause autodigestion?

A. Golgi bodies
B. Lysosomes
C. Microsomes
D. Peroxisomes

2-4. All the following activities are taking place inside mitochondria, *except:*

A. Citric acid cycle
B. Electron transport chain
C. Beta oxidation of fatty acid
D. De novo fatty acid synthesis

2-5. All the following statements are true with regard to biomembranes, *except:*

A. They are selectively permeable
B. Have interior hydrophilic and exterior hydrophobic groups
C. Lipid bilayers contain phospholipids and cholesterol
D. Carbohydrate residues are located on their surfaces

2-6. The fluid mosaic model of membrane includes:

A. Cholesterol molecules in the membrane increase stability of membrane
B. Membrane proteins are fixed and cannot move
C. Membrane contains bimolecular leaflet of lipids
D. Permeability of drugs across cell membrane is dependent on their solubility in water

2-7. All the following lipids are present in cell membranes, *except:*

A. Lecithin B. Cholesterol
C. Sphingomyelin D. Triacylglycerol

2-8. Fluidity of membranes depends on:

A. Nature of fatty acids
B. Concentration of proteins
C. Glycosylation of proteins
D. Membrane pumps

2-9. Sodium-potassium pump is inhibited by:

A. Phlorizin B. Digoxin
C. Valinomycin D. Dicoumarol

2-10. Which of the statements regarding active transport is *incorrect*?

A. Requires carrier protein
B. Energy dependent
C. Against concentration gradient
D. Transport of water is an example

2-11. All the following are channel formers, *except:*

A. Gramicidin B. Amelogenin
C. Valinomycin D. Glutamate

2-12. Symport system operates in the transport of:

A. Bicarbonate B. Chloride
C. Calcium D. Glucose

2-13. Glucose-6-phosphatase is a marker enzyme present in:

A. Cytoplasm B. Mitochondria
C. Lysosomes D. Microsomes

2-14. The fluid mosaic model of membrane includes:

A. Cholesterol molecules increase stability of membrane
B. Membrane proteins are fixed and cannot move
C. Membrane contains bimolecular leaflet of lipids
D. Permeability of drugs across cell membrane is dependent on their solubility in water

2-15. Galactosyl transferase is a marker enzyme of:

A. Cell membrane B. Mitochondria
C. Lysosomes D. Golgi complex

2-16. ATP synthase is a marker enzyme present in:

A. Cytoplasm B. Mitochondria
C. Lysosomes D. Golgi complex

2-17. All the following enzymes are present in lysosomes, *except:*
A. Alpha fucosidase
B. Aryl sulfatase
C. Beta glucuronidase
D. Superoxide dismutase

2-18. Golgi bodies are:
A. Involved in phagocytosis
B. Temporary storage sites for secretory products
C. Concerned with disposal of waste products
D. Engaged in transport of molecules into mitochondria

2-19. Which of the following are not synthesized in endoplasmic reticulum?
A. Lipoproteins
B. Glycoproteins
C. RNA
D. Ganglioside

2-20. Which of the diseases listed below is a lysosomal storage disease?
A. Zelleweger syndrome
B. Inclusion cell disease
C. Pompe's disease
D. Leber's neuropathy

2-21. Peroxisomes are:
A. The site of biological oxidation
B. The site of post-translational modifications
C. Concerned with drug detoxification
D. Involved in the production of free radicals

2-22. Which are specialized membrane structures?
A. Peroxisomes
B. Desmosomes
C. Lysosomes
D. Microsomes

2-23. All the following are channel formers, *except:*
A. Gramicidin
B. Amelogenin
C. Valinomycin
D. Glutamate

2-24. Phagocytosis involves:
A. Respiratory burst
B. Lipid peroxidation
C. Alpha oxidation
D. Endosmosis

2-25. Which of the following are features of endocytosis?
A. Plasma membrane is invaginated to form vesicles
B. Cytoplasmic side of vesicle is coated by clathrin filaments
C. LDL uptake is by endocytosis
D. Heparin helps in this mechanism

Answers (Chapter 2)

2-1. B.	2-2. B.	2-3. B.	2-4. D.
2-5. B.	2-6. C.	2-7. D.	2-8. A.
2-9. B.	2-10. D.	2-11. D.	2-12. B.
2.13. D.	2-14. C.	2-15. D.	2-16. B.
2-17. D.	2-18. B.	2-19. D	2-20. C.
2-21. D.	2-22. B.	2-23. D.	2-24. A.
2-25. D.			

CHAPTER 3: AMINO ACIDS

3-1. One of the amino acids listed below is not basic:
A. Arginine B. Histidine
C. Glutamine D. Lysine

3-2. Which of the following amino acid has a hydroxyl group?
A. Valine B. Threonine
C. Leucine D. Histidine

3-3. All the following amino acids are acidic, *except:*
A. Aspartic acid B. Tyrosine
C. Glutamic acid D. Threonine

3-4. All the following are branched chain amino acids, *except:*
A. Valine B. Leucine
C. Isoleucine D. Threonine

3-5. All the following are sulfur containing amino acids, *except:*
A. Cysteine B. Methionine
C. Homocysteine D. Threonine

3-6. Guanidinium group is present in:
A. Arginine
B. Tryptophan
C. Histidine
D. Proline

3-7. Indole ring is present in:
A. Arginine
B. Tryptophan
C. Histidine
D. Proline

3-8. The following amino acids have hydrophobic side chains, *except:*
A. Tyrosine
B. Lysine
C. Leucine
D. Valine

3-9. Imidazole ring is present in:
A. Arginine
B. Tryptophan
C. Histidine
D. Proline

3-10. Pyrrolidine group is present in which of the following amino acid?
A. Arginine
B. Tryptophan
C. Histidine
D. Proline

3-11. Which of the following amino acids has a non-polar side chain?
A. Serine
B. Valine
C. Asparagine
D. Threonine

3-12. All the following are glucogenic amino acids, *except:*
A. Glycine
B. Serine
C. Leucine
D. Aspartic acid

3-13. Which of the following amino acids has a net positive charge at physiological pH?
A. Glutamic acid
B. Lysine
C. Valine
D. Leucine

3-14. Which is the ketogenic amino acid?
A. Alanine
B. Glutamic acid
C. Leucine
D. Aspartic acid

3-15. All the following are essential amino acids, *except:*
A. Tyrosine
B. Lysine
C. Valine
D. Phenyl alanine

3-16. Non-essential amino acids:
A. Are not seen in tissue proteins
B. Could be synthesized in the body
C. Have no role in body metabolism
D. Have aromatic ring structure

3-17. At iso-electric pH, the amino acids and proteins show:

A. Maximum net charge
B. Maximum mobility in electric field
C. Maximum precipitability
D. Minimum buffering action

3-18. This amino acids has 2 optically active centers (asymmetric carbon atoms):

A. Glycine B. Isoleucine
C. Valine D. Tryptophan

3-19. Enzymes are activated by phosphorylation of which amino acid residue?

A. Cysteine B. Serine
C. Glutamic acid D. Lysine

3-20. The amino acid which has maximum buffering capacity at physiologic pH is:

A. Arginine B. Alanine
C. Histidine D. Glutamic acid

3-21. Which does not undergo decarboxylation?

A. Tyrosine B. S-adenosyl methionine
C. Dopamine D. Histidine

3-22. The protein which does not answer aldehyde test:

A. Hemoglobin B. Albumin
C. Casein D. Gelatin

Answers (Chapter 3)

3-1. C.	3-2. B.	3-3. D.	3-4. D.
3-5. D.	3-6. A.	3-7. B.	3-8. B.
3-9. C.	3-10. D.	3-11. B.	3-12. C.
3-13. B.	3-14. C.	3-15. A.	3-16. B.
3-17. C.	3-18. B.	3-19. B.	3-20. C.
3-21. C.	3-22. D.		

CHAPTER 4: PROTEINS

4-1. During denaturation, proteins will not lose their structure, with regard to:

A. Primary structure B. Secondary structure
C. Tertiary structure D. Quaternary structure

4-2. The characteristic features of the peptide bond include all the following, *except:*

A. Does not allow freedom of rotation
B. It is a partial double bond
C. Always has cis configuration
D. Absorbs UV light at 220 nm

4-3. The force maintaining the primary structure of a protein:

A. Peptide bonds
B. Hydrophobic forces
C. Hydrogen bonds
D. Electrostatic (ionic) bonds

4-4. The forces maintaining the secondary, tertiary and quaternary structures of a protein are the following, *except:*

A. Electrostatic (ionic) bonds
B. Hydrophobic forces
C. Vander Waal's forces
D. Peptide bonds

4-5. The amino acid which did not allow the formation of alpha-helix is:

A. Glutamate B. Proline
C. Tyrosine D. Histidine

4-6. Tertiary structure of a protein describes:

A. The sequence of amino acids
B. Location of disulphide bonds
C. Amino-terminal-end amino acid
D. The nature of protein folding

4-7. One of the following proteins does not have a quaternary structure:

A. Albumin
B. Hemoglobin
C. Lactate dehydrogenase
D. Immunoglobulin G

4-8. All the following reagents are used for identifying the first amino acid in a protein, *except*:

A. Cyanogen bromide
B. Fluorodinitro benzene
C. Dansyl chloride
D. Phenyl isothiocyanate

4-9. Proteins can be precipitated by the following methods, *except:*

A. Adding alcohol and acetone
B. Saturating with ammonium sulphate
C. Using salts of heavy metals
D. Shifting the pH away from the iso-electric point

4-10. Denatured proteins:

A. Are soluble
B. Are difficult to digest
C. Are biologically inactive
D. Peptide bonds are broken

4-11. Which of the following is a simple protein?

A. Casein B. Insulin
C. Hemoglobin D. Tyrosinase

4-12. In glycoproteins, the carbohydrate chains are combined through glycosidic linkages with:

A. Hydroxyl groups of serine or threonine residues of proteins
B. Epsilon amino nitrogen of lysine residues of proteins
C. Guanidium group of arginine residues of proteins
D. Phenol group of tyrosine residues of proteins

4-13. The protein which does not answer the aldehyde test is:

A. Hemoglobin B. Albumin
C. Casein D. Gelatin

4-14. Proteins may be estimated by the following methods, *except:*

A. Biuret method
B. Heat coagulation
C. Kjeldahl's digestion
D. Nephelometry

4-15. All the following are examples of tertiary and quaternary structures of proteins, *except:*

A. Alpha helix
B. Beta pleated sheet
C. Triple stranded helix
D. Peptide bonds

4-16. Tertiary structure of a protein describes:
A. Sequence of amino acids
B. Location of disulphide bonds
C. Amino-terminal-end amino acid
D. The nature of protein folding

4-17. Proteins may be denatured irreversibly by:
A. Adding urea
B. Bringing to iso-electric pH
C. Heat coagulation
D. Reduction with mercaptoethanol

4-18. Lectins are:
A. Animal proteins having specific amino acid binding site
B. Antibody molecules acting against cells
C. Plant proteins having specific carbohydrate binding site
D. Blood proteins having a lecithin group

4-19. Ultraviolet light at 280 nm is absorbed by which component of proteins?
A. Peptide bonds
B. Sulfhydryl group of cysteine
C. Indole ring of tryptophan
D. Imidazole ring of proline

Answers (Chapter 4)

4-1. A.	4-2. C.	4-3. A.	4-4. D.
4-5. B.	4-6. D.	4-7. A.	4-8. A.
4-9. D.	4-10. C.	4-11. B.	4-12. A.
4-13. D.	4-14. B.	4-15. D.	4-16. D
4-17. C.	4-18. D	4-19. C.	

CHAPTER 5: ENZYMOLOGY

5-1. An example of lyase is:
A. Glutamine synthetase
B. Fumarase
C. Cholinesterase
D. Amylase

5-2. The enzyme belonging to the class ligase is:
A. Glycogen synthase
B. Glutamine synthetase
C. Porphobilinogen deaminase
D. Histidine decarboxylase

5-3. Co-enzymes are:
A. Dialyzable, non-protein molecules
B. Colloidal protein molecules
C. Structural analogs of enzymes
D. Different forms of the same enzyme

5-4. Digestive enzymes belong to the class of:
A. Hydrolases B. Ligases
C. Lysases D. Oxidoreductases

5-5. All the following statements are true with regard to the enzymes called synthases and synthetases, *except:*
A. Synthetases and synthases are synonyms
B. Synthetases are enzymes belonging to ligases
C. Glutamine synthetase is belonging to class of ligases
D. Glycogen synthase belongs to the class transferases

5-6. The co-enzyme:
A. Increases the affinity of apo-enzyme to substrate
B. Lowers the activation energy
C. Increases the number of active sites on the enzyme
D. Accepts one of the products of the reaction

5-7. All the following metalloenzymes contain copper, *except*:
A. Superoxide dismutase
B. Tyrosinase
C. Glutathione peroxidase
D. Cytochrome oxidase

5-8. Which one of the following enzymes is containing molybdenum?
A. Cytochrome oxidase
B. Xanthine oxidase
C. Carbonic anhydrase
D. Phosphoglucomutase

5-9. The enzyme:
A. Reduces the energy of activation
B. Increases total energy of substrate
C. Increases the equilibrium constant
D. Increases total energy of the product

5-10. In enzyme kinetics, Vmax denotes:
A. The amount of an active enzyme
B. Substrate concentration
C. Half the substrate concentration
D. Quantity of enzyme substrate complex

5-11. All the following statements are true with regard to enzymes, *except:*
A. Enzymes lower activation energy
B. Thoy altor oquilibrium of tho roaction
C. They accelerate the chemical reaction
D. Most of the enzymes are proteins in nature

5-12. The Km value of an enzyme is:
A. The substrate concentration at half maximal velocity
B. Half the substrate concentration at maximal velocity
C. Dissociation constant of enzyme-substrate complex
D. The total enzyme concentration

5-13. Lineweaver-Burk (double reciprocal) graph is plotted by:
A. Substrate concentration against Vmax
B. Reciprocal of substrate concentration against reciprocal of velocity
C. Reciprocal of initial velocity against substrate concentration
D. Km against initial velocity

5-14. When substrate concentration is equal to Km value:
A. Half of the enzyme molecules are bound to the substrate molecules and the other half are free
B. Maximum velocity is achieved
C. Maximum enzyme molecules are taking part in the reaction
D. The reaction is now at equilibrium

5-15. Enzymes which are synthesized in inactive form are called:

A. Co-enzymes B. Apo-enzymes
C. Lysozymes D. Pro-enzymes

5-16. All the following are examples of competitive inhibition, *except:*

A. Inhibition of enolase by fluoride ions
B. Inhibition of dihydrofolate reductase by amethoptrine
C. Treatment of methanol poisoning by ethanol
D. Use of dicoumarol as an anticoagulant

5-17. In competitive inhibition:

A. Inhibitor has structural similarity to substrate
B. Km is decreased
C. Vmax is decreased
D. Reaction rate is independent of substrate concentration

5-18. In a competitive inhibition:

A. Km is increased and Vmax is increased
B. Km is decreased and Vmax is normal
C. Km is increased and Vmax is normal
D. Km is decreased and Vmax is increased

5-19. The antimetabolite, methotrexate acts by:

A. Non-competitive inhibition
B. Formation of a covalent bond between enzyme and inhibitor
C. Increasing the Vmax of the reaction
D. Competitive inhibition

5-20. Ascaris worm is not digested by intestinal enzymes, because:

A. The worm produces a trypsin inhibitor
B. Ascaris protein stimulates digestive enzymes
C. Ascaris has a tough cuticle
D. Proteins of the worm is resistant to trypsin

5-21. The following are salient features of allosteric enzymes, *except:*

A. They are usually multiple subunit enzymes
B. They are involved in metabolic regulation
C. They do not show any change in Km and Vmax in the presence of modifiers
D. On binding the inhibitor, usually Km is increased and Vmax is decreased

5-22. Allosteric enzymes show all the following characteristics, *except:*

A. Cooperative binding of the substrate
B. Sigmoid kinetics
C. Binding between substrate and regulatory sites
D. Substrate binding site and regulatory site are different

5-23. All the following enzymes are examples of covalent modifications, *except:*

A. Chymotrypsinogen
B. Pyruvate kinase
C. Glycogen phosphorylase
D. Aspartate transcarbamoylase

5-24. All are multienzyme complexes, *except:*

A. Fatty acid synthase
B. Pyruvate dehydrogenase
C. Alpha ketoglutarate dehydrogenase
D. Glycogen synthase

5-25. One of the following is an example of repression:

A. ALA synthase by heme
B. Succinate dehydrogenase by malonate
C. Cytochromes by cyanide
D. Glycogen phosphorylase by glucagon

5-26. All the following are examples of enzyme induction, *except:*

A. Beta galactosidase by lactose
B. Tryptophan pyrrolase by glucocorticoid
C. Transaminases by insulin
D. ALA synthase by barbiturates

5-27. Which enzyme requires chloride ion as activator?

A. Amylase B. Lipase
C. Trypsin D. Elastin

5-28. All compounds listed below act as competitive inhibitors of enzymes, *except:*

A. Methotrexate
B. Sulfonamide
C. Allopurinol
D. Fluoride

5-29. A competitive inhibitor will affect the enzyme activity by:
A. Precipitating the enzyme protein
B. Decreasing Km value
C. Increasing Vmax
D. Irreversibly denaturing the enzyme

5-30. In an enzyme assay, the substrate concentration is lower than the Km, then the rate is:
A. Independent of enzyme concentration
B. Independent of temperature
C. Proportionate to the substrate concentration
D. Approximately equal to Vmax.

5-31. Enzyme activity in biological systems may be affected by all the following, *except:*
A. Substrate analogs
B. Inducers
C. Immobilization of enzymes
D. Diisopropylfluoro phosphate

5-32. Iodoacetate inhibits enzyme by reacting with which particular group?
A. Amide B. Sulfhydryl
C. Carboxyl D. Imidazole

5-33. Enzymes synthesized in inactive form are called:
A. Co-enzymes B. Apo-enzymes
C. Lysozymes D. Pro-enzymes

5-34. A constitutive enzyme shows all the features, *except:*
A. Activated by an inducer
B. Inhibited by a repressor
C. Always present in the cell in the same concentration
D. A constituent of the cell membrane

5-35. Synthesis of an enzyme is increased by its substrate; this phenomenon is called as:
A. Allosteric activation
B. Induction
C. Cooperative activation
D. Feed forward activation

Answers (Chapter 5)

5-1. B.	5-2. B.	5-3. A.	5-4. A.
5-5. A.	5-6. D.	5-7. C.	5-8. B.
5-9. A.	5-10. A.	5-11. B.	5-12. A.
5-13. B.	5-14. A.	5-15. D.	5-16. A.
5-17. A.	5-18. C.	5-19. D.	5-20. A.
5-21. C.	5-22. C.	5-23. D.	5-24. D.
5-25. A.	5-26. C.	5-27. A.	5-28. D.
5-29. D.	5-30. C.	5-31. C.	5-32. B.
5-33. D.	5-34. C.	5-35. B.	

CHAPTER 6: CHEMISTRY OF CARBOHYDRATES

6-1. Glucose and galactose are:
A. Anomers B. Constituents of sucrose
C. Diasterio isomers D. Epimers

6-2. D-mannose is:
A. The 4th epimer of glucose
B. A keto sugar
C. Is a component of maltose
D. Is the 2nd epimer of glucose

6-3. A pair of sugars differing from each other in the functional group is called:
A. Anomers B. Epimers
C. Racemers D. Stereo isomers

6-4. Reducing sugars are differentiated from non-reducing sugars by:
A. Iodine test B. Molisch's test
C. Seliwanoff's test D. Benedict's test

6-5. In the case of monosaccharides, all properties go hand in hand, *except:*
A. Reducing property
B. Formation of furfural
C. Formation of osazone
D. Mutarotation

6-6. All the following are sugar alcohols, *except:*
A. Dulcitol B. Mannitol
C. Xylulose D. Sorbitol

6-7. The glycosidic linkage seen in sucrose is:
A. Alpha 1-4 linkage B. Beta 1-4 linkage
C. Alpha 1-6 linkage D. 1-2 linkage

6-8. All the following have glycosidic bond, *except:*
A. Maltose
B. Sucrose
C. N-acetyl glucosamine
D. Alpha methyl glucose

6-9. The sugar found in milk is:
A. Galactose B. Glucose
C. Fructose D. Lactose

6-10. Sucrose consists of:
A. Glucose + glucose
B. Glucose + fructose
C. Glucose + galactose
D. Glucose + mannose

6-11. Which is a non-reducing sugar?
A. Maltose B. Sucrose
C. Lactose D. Isomaltose

6-12. The glycosidic linkage seen in lactose is:
A. Alpha 1-4 linkage B. Beta 1-4 linkage
C. Alpha 1-6 linkage D. Alpha 1-2 linkage

6-13. The glycosidic linkage seen in maltose is:
A. Alpha 1-4 linkage B. Beta 1-4 linkage
C. Alpha 1-6 linkage D. Alpha 1-2 linkage

6-14. Hydrolysis of maltose will give rise to:
A. Glucose only
B. Glucose and fructose
C. Glucose and galactose
D. Glucose and mannose

6-15. Which contains alpha glycoside linkages?
A. Chondroitin sulfate
B. Keratan sulfate
C. Hyaluronic acid
D. Amylopectin

6-16. Which contains a beta glycosidic linkage?
A. Heparin B. Glycogen
C. Cellulose D. Starch

6-17. All the following are homopolysaccharides, *except:*
A. Chitin B. Glycogen
C. Inulin D. Heparin

6-18. The heteropolysaccharide which does not contain uronic acid is:
A. Keratan sulfate
B. Dermatan sulfate
C. Chondroitin sulfate
D. Hyaluronic acid

6-19. On oxidation, glucose will give rise to all the following, *except:*
A. Glucuronic acid
B. Gluconic acid
C. Glucosaccharic acid
D. Glycoside

6-20. Hyaluronic acid is seen in:
A. Synovial fluid
B. Cornea of eye
C. Outer shells of insects
D. Cartilage

6-21. Beta glycosidic bond is present in:
A. Starch B. Heparin
C. Lactose D. Glycogen

Answers (Chapter 6)

6-1. D.	6-2. D.	6-3. A.	6-4. D.
6-5. B.	6-6. C.	6-7. D.	6-8. C.
6-9. D.	6-10. B.	6-11. B.	6-12. B.
6-13. A.	6-14. A.	6-15. D.	6-16. C.
6-17. D.	6-18. A.	6-19. D.	6-20. A.
6-21. C			

CHAPTER 7: CHEMISTRY OF LIPIDS

7-1. Which of the following is not a phospholipid?
A. Sphingomyelin B. Cerebroside
C. Cephalin D. Lecithin

7-2. The structure of stearic acid is:
A. 16 carbon, no double bond
B. 18 carbon, no double bond
C. 18 carbon, two double bonds
D. 18 carbon, three double bonds

7-3. The structure of oleic acid can be described as:
A. 16 carbon, one double bond
B. 18 carbon, one double bond
C. 18 carbon, two double bonds
D. 18 carbon, three double bonds

7-4. The structure of linolenic acid is:
A. 16 carbon, one double bond
B. 18 carbon, one double bond
C. 18 carbon, two double bonds
D. 18 carbon, three double bonds

7-5. The following are true with regard to linolenic acid, *except:*
A. It is a component of lecithin
B. It is present in sunflower oil
C. It can be synthesized by liver
D. It has three double bonds

7-6. The major fat in adipose tissue is:
A. Phospholipid B. Cholesterol
C. Sphingolipids D. Triacyl glycerol

7-7. Which of the following lipids is not present in vegetable oils?
A. Cholesterol B. Linoleic acid
C. Oleic acid D. Stearic acid

7-8. Neutral fats are:
A. Act as surfactant and detergents
B. Soluble in water
C. Components of biomembranes
D. Oxidized to give energy

7-9. Saponification number of a fat molecule:
A. Decreases with increase in molecular weight of fat
B. Increases with increase in molecular weight of fat
C. Decreases with increase in number of double bonds
D. Increases with increase in number of double bonds

7-10. The nitrogenous base present in lecithin is:
A. Choline B. Ethanolamine
C. Serine D. Sphingosine

7-11. All the following statements are true with regard to phospholipids, *except:*
A. They can exist as zwitter ions
B. They have surfactant properties
C. They are components of biomembranes
D. They are resistant to the action of enzymes

7-12. All the following alcohols are found in phospholipids, *except:*
A. Sphingosine B. Inositol
C. Mannitol D. Glycerol

7-13. Sphingomyelin on hydrolysis yields all the following, *except:*
A. Sphingosine B. Glucose
C. Phosphate D. Choline

7-14. Cardiolipin:
A. Contains choline
B. Is diphosphatidyl glycerol
C. Is a part of sphingolipin
D. Is found in endoplasmic reticulum

7-15. A ganglioside on hydrolysis gives all the following, *except:*
A. Fatty acid B. Glycerol
C. Sphingosine D. N-acetyl neuraminic acid

7-16. Hydrolysis of fat by alkali is called:
A. Esterification B. Saponification
C. Mutarotation D. Alkylation

7-17. Which one of the following is a dietary essential?
A. Oleic acid B. Palmitic acid
C. Stearic acid D. Linolenic acid

7-18. Which of the following lipids give 2 fatty acids, one molecule of glycerol and one molecule of phosphoric acid on complete hydrolysis?
A. Diacyl glycerol B. Phosphatidic acid
C. Lecithin D. Cephalin

7-19. Which of the following fatty acids is present in surfactant ?
A. Palmitic acid
B. Stearic acid
C. Homogamma linolenic acid
D. Arachidonic acid

7-20. Which of the following is not an amphipathic molecule?
A. Sodium glycocholate
B. Cholesterol linoleate
C. Phosphatidyl serine
D. Potassium taurocholate

Answers (Chapter 7)

7-1. B.	7-2. B.	7-3. B.	7-4. D.
7-5. C.	7-6. D.	7-7. A.	7-8. D.
7-9. A.	7-10. A.	7-11. D.	7-12. C.
7-13. B.	7-14. B.	7-15. B.	7-16. B.
7-17. D.	7-18. B.	7-19. A	7-20. B.

CHAPTER 9: MAJOR PATHWAYS OF GLUCOSE

9-1. Glucose is absorbed by gastrointestinal cells by:
A. Passive diffusion
B. Carrier mediated co-transport system, along with sodium
C. Carrier mediated antiport system, along with potassium
D. Needs no ATP

9-2. The product of glycolysis in erythrocytes is:
A. NADPH B. Lactate
C. Pyruvate D. Carbon dioxide

9-3. Glucokinase is more active after a meal, because:
A. It is an inducible enzyme
B. It has more affinity to glucose than hexokinase
C. It is present in all tissues
D. Can act on all monosaccharides

9-4. Which enzyme catalyzes an irreversible reaction?
A. Transketolase
B. Phosphofructokinase
C. Aldolase
D. Glyceraldehyde-3-phosphate dehydrogenase

9-5. An example of substrate level phosphorylation is:

A. Phosphoglycerate kinase
B. Enolase
C. Pyruvate kinase
D. Glyceraldehyde-3-phosphate dehydrogenase

9-6. All the dehydrogenases listed below are NAD^+ dependent, *except:*

A. Lactate dehydrogenase
B. Glucose-6-phosphate dehydrogenase
C. Pyruvate dehydrogenase
D. Glyceraldehyde-3-phosphate dehydrogenase

9-7. In which enzymes, glycolytic pathway is inhibited by fluoride ions?

A. Hexokinase
B. Phospho fructo kinase
C. Aldolase
D. Enolase

9-8. As the end product of glycolysis, pyruvate and NADH are formed. During anaerobiosis, the NAD^+ is regenerated by:

A. Glyceraldehyde-3-phosphate dehydrogenase
B. Oxygen
C. Glutamate dehydrogenase
D. Lactate dehydrogenase

9-9. As the end product of glycolysis, pyruvate and NADH are formed. During aerobic conditions, the NAD^+ is regenerated by:

A. Glyceraldehyde-3-phosphate dehydrogenase
B. Oxygen
C. Glutamate dehydrogenase
D. Lactate dehydrogenase

9-10. All the enzymes listed below are regulatory enzymes, *except:*

A. Glycogen phosphorylase
B. Glucose-6-phosphate dehydrogenase
C. Pyruvate kinase
D. Lactate dehydrogenase

9-11. ATP is generated in all the reactions *except:*

A. Glyceraldehyde-3-phosphate dehydrogenase
B. 1,3-bis phosphoglycerate kinase
C. Pyruvate kinase
D. Hexokinase

9-12. The key enzyme of glycolysis is:
A. Glucose-6-phosphatase
B. Glyceraldehyde-3-phosphate dehydrogenase
C. Phospho hexose isomerase
D. Phosphofructokinase

9-13. Catalytic activity of phosphofructokinase is increased by all the following, *except:*
A. AMP
B. Fructose-2, 6-bis phosphate
C. ATP
D. Fructose-6-phosphate

9-14. Complete oxidation of one molecule of glucose yields how many ATPs?
A.12 B. 24
C. 32 D. 129

9-15. Rapoport-Leubering Cycle in RBC produces:
A. More ATP than produced in Embden-Meyerhof pathway
B. More NADPH than HMP shunt pathway
C. 2,3-BPG which enhances oxygen dissociation
D. 1,3-BPG which contains a high energy bond

9-16. All the following co-enzymes are involved in the pyruvate dehydrogenase reaction, *except:*
A. Thiamine pyrophosphate (TPP)
B. Biotin
C. NAD^+
D. FAD

9-17. There is no net synthesis of glucose from fatty acids, because of the irreversible nature of the reaction:
A. Pyruvate to oxaloacetate
B. Pyruvate to acetyl CoA
C. Phosphoenol pyruvate to pyruvate
D. Oxaloacetate to phosphoenol pyruvate

9-18. Pyruvate is converted to acetyl CoA by:
A. Pyruvate dehydrogenase
B. Pyruvate carboxylase
C. Pyruvate kinase
D. Lactate dehydrogenase

9-19. Malate shuttle is necessary for:
A. Gluconeogenesis B. Providing NADPH
C. ATP generation D. Glycogenolysis

9-20. Gluconeogenesis is inhibited by:
A. Glucagon B. Growth hormone
C. Insulin D. Glucocorticoids

9-21. All the following are key gluconeogenic enzymes, *except:*
A. Pyruvate carboxylase
B. Phosphoenol pyruvate carboxy kinase
C. Phosphofructokinase
D. Glucose-6-phosphatase

9-22. Blood glucose level can be raised by gluconeogenesis only by liver because:
A. Glucokinase is present exclusively in liver
B. Glucose-6-phosphatase is active only in liver
C. Transaminases are present in liver
D. Liver is the site of storage of glucose as glycogen

9-23. How many ATP molecules are required to convert 2 molecules of pyruvate into glucose?
A. Two B. Three
C. Six D. Eight

9-24. All the following are substrates for gluconeogenesis, *except:*
A. Palmitic acid B. Lactic acid
C. Alanine D. Glycerol

9-25. All the following statements are true with regard to glucokinase, *except:*
A. It has lower affinity for glucose than hexokinase
B. It is found only in liver
C. It is induced by insulin
D. It phosphorylates glucose under conditions of fasting

9-26. Which glycolytic enzyme uses inorganic phosphate for the reaction?
A. Pyruvate kinase
B. 1,3-bis phospho glycerate kinase
C. Glyceraldehyde-3-phosphate dehydrogenase
D. Hexokinase

9-27. All will inhibit phosphofructokinase, *except:*
A. Fructose-2,6-bis phosphate
B. ATP
C. Cyclic AMP
D. Citrate

9-28. 2,3-bis phosphoglycerate is:
A. Binding to fetal hemoglobin
B. A normal intermediate of glycolytic pathway
C. A high energy compound
D. Needed for triglyceride formation

9-29. All the following statements are true with regard to gluconeogenesis, *except:*
A. It maintains blood glucose level during fasting
B. Allows use of proteins as an alternate source of energy
C. There is net conversion of acetyl CoA to glucose during gluconeogenesis
D. Glycerol part of fat forms a minor source for gluconeogenesis

9-30. Pyruvate can be directly converted to all the following, *except:*
A. Alanine B. Lactate
C. Oxaloacetate D. Phosphoenol pyruvate

9-31. Gluconeogenesis is taking place in:
A. Brain B. Liver
C. RBC D. Muscle

9-32. Both glycolysis and gluconeogenesis pathways utilize all the following enzymes, *except:*
A. Phosphofructokinase
B. Glyceraldehyde-3-phosphate dehydrogenase
C. 1,3-bis phospho glycerate kinase
D. Phosphoglucomutase

9-33. Lactic acidosis is seen in the following cases, *except:*
A. OXPHOS diseases
B. Defective metabolism of pyruvate
C. Excessive ingestion of alcohol
D. Deficiency of glucose-6-phosphate dehydrogenase

9-34. Which of the following metabolic intermediates is not generated from glucose?
A. Dihydroxy acetone phosphate
B. 2,3- bisphospho glycerate
C. Beta hydroxy butyrate
D. Glycerol-3-phosphate

9-35. Which reactions does not occur in the liver?
A. Acetoacetate to acetoacetyl CoA
B. Hydroxylation of cholecalciferol
C. Glucose to UDP glucose
D. Oxaloacetate to phosphoenol pyruvate

9-36. Transport of glucose from intestinal lumen to mucosal cell:
A. Requires hydrolysis of ATP
B. Occurs by facilitated diffusion
C. Is sodium dependent
D. Occurs by an antiport system

Answers (Chapter 9)

9-1. B.	9-2. B.	9-3. A.	9-4. B.
9-5. D.	9-6. B.	9-7. D.	9-8. D.
9-9. B.	9-10. D.	9-11. D.	9-12. D.
9-13. C.	9-14. C.	9-15. C.	9-16. B.
9-17. B.	9-18. A.	9-19. A.	9-20. C.
9-21. C.	9-22. B.	9-23. C.	9-24. A.
9-25. D.	9-26. C.	9-27. A.	9-28. C.
9-29. C.	9-30. D.	9-31. B.	9-32. A.
9-33. D	9-34. C	9-35. D	9-36. C

CHAPTER 10: OTHER PATHWAYS OF GLUCOSE

10-1. Which enzyme generates NADPH?
A. Pyruvate carboxylase
B. Pyruvate dehydrogenase
C. Glucose-6-phosphate dehydrogenase
D. Lactate dehydrogenase

10-2. Transketolase activity is decreased in the deficiency of:
A. Thiamine pyrophosphate (TPP)
B. Nicotinamide adenine dinucleotide
C. Flavin adenine dinucleotide
D. Pyridoxal phosphate

10-3. Deficiency of glucose-6-phosphate dehydrogenase causes:
A. Cataract B. Hemolytic anemia
C. Hypoglycemia D. Mental retardardation

10-4. The HMP shunt pathway is important for all the following, *except:*
A. Generation of ATP
B. Fatty acid biosynthesis
C. Synthesis of reduced glutathione
D. Synthesis of ribose

10-5. All are true with regard to NADPH, *except:*
A. Generated by glucose-6-phosphate dehydrogenase
B. Used for reductive biosynthesis of fatty acids
C. Produces peroxides and free radicals in RBC
D. Preserves integrity of RBC membrane

10-6. The hormone activating the enzyme phosphorylase is:
A. Epinephrine B. Insulin
C. Growth hormone D. Glucocorticoids

10-7. Von-Gierke's disease is characterized by the deficiency of:
A. Glucose-6-phosphatase
B. Glyceraldehyde-3-phosphate dehydrogenase
C. Phosphofructokinase
D. Phosphorylase

10-8. Mc Ardle's syndrome is due to the deficiency of:
A. Glucose-6-phosphatase
B. Glucose-6-phosphate dehydrogenase
C. Muscle glycogen phosphorylase
D. Debranching enzyme

10-9. Glycogen is:
A. Having unbranched or straight line structure
B. Hydrolyzed to equal quantity of glucose and galactose
C. Hydrolysed by phosphorylase
D. Seen in brain

10-10. One glycosyl residue in glycogen is converted to 2 lactic acid molecules, the net yield of ATP is:
A. Two B. Three
C. Four D. Thirty eight

10-11. Which enzyme is active in the phosphorylated state:

A. Glycogen phosphorylase
B. Pyruvate kinase (PK)
C. PFK-2 (phospho fructo kinase-2)
D. Glycogen synthase

10-12. Glycogen synthase is activated by:

A. Insulin B. Protein kinase
C. Adrenaline D. ATP

10-13. Which enzyme uses NADP as co-enzyme?

A. Pyruvate carboxylase
B. Pyruvate dehydrogenase
C. Glucose-6-phosphate dehydrogenase
D. Lactate dehydrogenase

10-14. Glucose-6-phosphate dehydrogenase enzyme:

A. Is a member of glycolytic pathway
B. Is necessary for production of NADPH
C. Requires the co-enzyme ATP
D. Is inhibited by insulin

10-15. Transketolase reaction:

A. Takes place between xylulose phosphate and ribose phosphate
B. Occurs between glyceraldehyde phosphate and fructose phosphate
C. Needs pyridoxal phosphate as co-enzyme
D. Occurs between glyceraldehyde phosphate and sedoheptulose phosphate

10-16. Hemolytic episode after administration of antimalarial drug is due to the deficiency of enzyme:

A. 6-phosphogluconate dehydrogenase
B. Glucose-6-phosphate dehydrogenase
C. Glucose-6-phosphatase
D. Hexokinase

Answers (Chapter 10)

10-1. C.	10-2. A.	10-3. B.	10-4. A.
10-5. C.	10-6. A.	10-7. A.	10-8. C.
10-9. D	10-10. B.	10-11. A.	10-12. A.
10-13. C.	10-14. B.	10-15. A.	10-16. B.

CHAPTER 11: METABOLISM OF OTHER CARBOHYDRATES

11-1. All are true with regard to fructose intolerance, *except:*
A. Defective enzyme is aldolase-B
B. Fructose-1-phosphate accumulates
C. Glycogen phosphorylase is inhibited
D. Urine is positive for Rothera's test

11-2. All are correct with regard to fructose metabolism, *except:*
A. Fructose is metabolized in liver
B. Free fructose is seen in seminal plasma
C. Block in seminal vessels is indicated by the absence of fructose in semen
D. Compared to glucose, fructose is sluggishly metabolized

11-3. Features of galactosemia include the following, *except:*
A. Cataract
B. Hepatosplenomegaly
C. Mental retardation
D. Hemolytic anemia

11-4. Which inborn error will result in mental retardation?
A. Fructosuria
B. Essential pentosuria
C. Galactosemia
D. Glucose-6-phosphate dehydrogenase deficiency

11-5. Excess intake of alcohol may produce lactic acidosis, because:
A. Alcohol is oxidized to lactic acid
B. Alcohol is oxidized to acetaldehyde and then to lactic acid
C. NADH is generated, which converts pyruvate to lactate
D. Ethanol induces Cori's cycle, by which lactic acid produced from muscle, is taken to liver

11-6. N-acetyl neuraminic acid (NANA) (sialic acid) is synthesized by:

A. N-acetyl glucosamine + glyceraldehyde-3-phosphate
B. Galactose-6-phosphate + pyruvate
C. Glucose-6-phosphate + lactate
D. N-acetyl mannosamine-6-phosphate + phosphoenol pyruvate

11-7. Mannose derivatives are produced from:

A. Glucosamine
B. N-acetyl glucosamine
C. Glucose-1-phosphate
D. UDP-glucose

11-8. Hereditary fructose intolerance is usually manifested when the child is about 6 months old, because at that age:

A. Defective enzyme is produced in liver
B. Fructose is introduced in diet
C. Mother's milk is reduced
D. Starch diet is being introduced

11-9. Fructokinase catalyzes which reaction?

A. Fructose ⟶ fructose-6-phosphate
B. Fructose ⟶ fructose-1-phosphate
C. Fructose ⟶ fructose-1,6-bisphosphate
D. Fructose ⟶ fructose-2,6-bisphosphate

11-10. Which is true with regard to fructosuria?

A. It is due to the absence of Aldolase-B
B. Severe mental retardation is seen
C. Urine is positive for Benedict's test
D. Glycolysis is inhibited

11-11. In glycoproteins, the outermost carbohydrate residue is donated by:

A. ADP-xylose B. UDP-glucosamine
C. CMP-NANA D. GDP-fucose

11-12. Cataracts are formed due to accumulation of:

A. Xylitol in essential pentosuria
B. Galactitol (dulcitol) in galactosemia
C. Mannitol in galactosemia
D. Ribitol in renal glycosuria

11-13. Excess ingestion of alcohol may produce hypoglycemia, because ethanol:
A. Inhibits gluconeogenesis
B. Favors glycogen synthesis
C. Increases secretion of insulin
D. Inhibits absorption of glucose

Answers (Chapter 11)

11-1. D.	11-2. D.	11-3. D.	11-4. C.
11-5. C.	11-6. D.	11-7. B.	11-8. B.
11-9. B.	11-10. C.	11-11. C.	11-12. B.
11-13. A.			

CHAPTER 12: METABOLISM OF FATTY ACIDS

12-1. The complete digestion of triacyl glycerol (triglyceride) in the gastrointestinal tract needs all the following, *except:*
A. Pancreatic lipase B. Hormone sensitive lipase
C. Isomerase D. Bile salts

12-2. Bile salts help in the absorption of dietary lipids by:
A. Producing the micellar state of lipids
B. Incorporating cholesterol into chylomicrones
C. Converting triglycerides into monoglycerides
D. Providing the optimum pH for lipase activity

12-3. All are correct with regard to carnitine, *except:*
A. It transports fatty acids from cytoplasm to mitochondria
B. Carnitine is involved in fatty acid synthesis
C. Medium chain fatty acids do not require carnitine
D. Carnitine is synthesized from lysine

12-4. During each cycle of beta oxidation of fatty acids, all the following are generated, *except:*
A. Fatty acyl CoA, with 2 carbon atoms less
B. Acetyl CoA
C. $FADH_2$
D. NADPH

12-5. The net gain of ATP per molecule of palmitic acid on complete oxidation is:

A. 12 B. 38
C. 106 D. 135

12-6. The co-enzyme required for propionyl CoA carboxylase is:

A. Thiamine pyrophosphate
B. NADPH
C. Biotin
D. Pyridoxal phosphate

12-7. Name the defective enzyme in propionic acidemia:

A. Propionyl CoA carboxylase
B. Methyl malonyl CoA racemase
C. Tyrosinase
D. Propionic acid dehydrogenase

12-8. Name the defective enzyme in methyl malonyl aciduria:

A. Propionyl CoA carboxylase
B. Methyl malonyl CoA mutase
C. Tyrosinase
D. Cystathionine synthase

12-9. Refsum's disease is due to lack of enzymes of:

A. Beta oxidation B. Omega oxidation
C. Alpha oxidation D. Desaturation

12-10. Refsum's disease is due to accumulation of:

A. Phytanic acid B. Sphingomyelin
C. Glucocerebroside D. Ganglioside

12-11. Acetyl CoA formed in the mitochondria is made available in the cytoplasm for de novo synthesis of fatty acids by the activity of the enzyme:

A. Carnitine acyl transferase
B. ATP citrate lyase
C. Acetyl CoA carboxylase
D. Pyruvate dehydrogenase

12-12. The key enzyme in fatty acid synthesis is:

A. Acetyl CoA carboxylase
B. Beta hydroxy acyl dehydratase
C. Enoyl reductase
D. Acetyl transacylase

12-13. Acetyl CoA carboxylase is activated by:
A. Palmityl CoA B. Carbon dioxide
C. Citrate D. Malonyl CoA

12-14. For the de novo fatty acid synthesis, the co-enzyme required is:
A. NADH
B. NADPH
C. Thiamine pyrophosphate
D. $FADH_2$

12-15. Sources of NADPH for fatty acid synthesis include all the following reactions, *except*:
A. Glucose-6-phosphate dehydrogenase
B. 6-phospho gluconate dehydrogenase
C. Cytoplasmic malate dehydrogenase
D. Cytoplasmic isocitrate dehydrogenase

12-16. In the biosynthesis of triacylglycerol from glycerol phosphate, all the following are intermediates, *except*:
A. Lysophosphatidic acid
B. Phosphatidic acid
C. 1,2-diacyl glycerol
D. Monoacyl glycerol

12-17. In adipose tissue, activity of hormone sensitive lipase is increased by all the following hormones, *except:*
A. Adrenalin B. Growth hormone
C. Corticosteroids D. Insulin

12-18. Chain elongation of fatty acids takes place in which of the following subcellular fraction?
A. Cytoplasm B. Nucleus
C. Microsomes D. Golgi complex

12-19. HMG CoA is a precursor of all the following, *except:*
A. Arachidonic acid B. Cholesterol
C. Acetone D. Beta hydroxy butyrate

12-20. HMG CoA is directly converted to all the following, *except*:
A. Acetoacetyl CoA B. Mevalonate
C. Acetoacetate D. Acetyl CoA

12-21. Fatty liver may be prevented by the following, *except:*

A. Ethanol
B. Choline
C. Methionine
D. Lecithin

12-22. Ketone bodies are produced mainly in:

A. Brain
B. Liver
C. Erythrocytes
D. Skeletal muscles

12-23. Ketosis is due to increased:

A. Hepatic glucose liberation
B. Carbohydrate utilization
C. Gluconeogenesis
D. Fatty acid oxidation

12-24. Which of the following tissues are capable of oxidizing ketone bodies?

A. Liver
B. Heart
C. RBCs
D. Plasma

12-25. Ketone bodies in urine are identified by:

A. Heat and acetic acid test
B. Rothera's test
C. Benzidine test
D. Benedict's test

12-26. Beta oxidation of fatty acids yields all the following, *except*:

A. NADH
B. Acetyl CoA
C. $FADH_2$
D. Malonyl CoA

12-27. Regarding propionic acid metabolism:

A. Propionyl CoA carboxylase is biotin dependent
B. Methyl malonyl CoA racemase is vitamin B_{12} dependent
C. Succinyl CoA is converted to propionyl CoA
D. Leucine is catabolized to produce propionyl CoA

12-28. All the enzymes are required for de novo synthesis of fatty acid, *except:*

A. Acetyl CoA carboxylase
B. Glucose-6-phosphate dehydrogenase
C. Thiolase
D. Malonyl transacylase

12-29. The action of caffeine include all the following, *except*:

A. Inhibition of phosphodiesterase
B. Prolongation of action of cyclic AMP
D. Increase in activity of hormone sensitive lipase
C. Inhibition of synthesis of prostaglandins

12-30. Ketosis is often seen in:

A. Nephritis
B. Coronary artery disease
C. Muscle degeneration
D. Diabetes mellitus

12-31. The pathway which takes place in the mitochondria is:

A. Ketogenesis
B. Glycolysis
C. HMP shunt pathway
D. Cholesterol synthesis

12-32. Utilization of ketone bodies by peripheral tissues needs:

A. NADPH dependent dehydrogenase
B. Biotin dependent carbon dioxide fixation reaction
C. Carbon dioxide elimination reaction
D. Succinyl CoA dependent thiophorase reaction

12-33. All are true with regard to ketone bodies *except*:

A. Acetoacetate, acetone and beta hydroxy butyrate
B. Seen in urine in uncontrolled diabetes mellitus
C. Mainly produced in adipose tissue
D. Seen in urine during prolonged starvation

12-34. Defects in oxidation of fatty acids can produce all *except*:

A. Hypoglycemia B. Ketonemia
C. Myopathy D. Hyperammonemia

12-35. Defective digestion and absorption of fats can lead to the following, *except*:

A. Steatorrhea
B. Hypocalcemia
C. Defective coagulation
D. Difficulty in dim light

12-36. Which of the following tissues cannot derive energy from fatty acids?

A. Brain B. Cardiac muscle
C. Skeletal muscle D. Erythrocytes

12-37. Which of the following processes does not occur in adipose tissue?

A. Formation of triose phosphates
B. Generation of NADPH
C. Hydrolysis of triacylglycerol
D. Phosphorylation of glycerol

12-38. The enzyme which is active only in the liver is:

A. HMG CoA reductase
B. Carnitine acyl transferase
C. Lecithin cholesterol acyl transferase
D. HMG CoA lyase

12-39. Fatty acid synthesis differs from beta oxidation in all, *except*:

A. Uses NADPH as reducing power
B. Requires co-enzyme A
C. Catalyzed by multienzyme complex
D. Activated by insulin

Answers (Chapter 12)

12-1. B.	12-2. A.	12-3. B.	12-4. D.
12-5. C.	12-6. C.	12-7. A.	12-8. B.
12-9. C.	12-10. A.	12-11. B.	12-12. A.
12-13. C.	12-14. B.	12-15. D.	12-16. D.
12-17. D.	12-18. C.	12-19. A.	12-20. A
12-21. A.	12-22. B.	12-23. D.	12-24. B
12-25. B.	12-26. D.	12-27. A.	12-28. C
12-29. C.	12-30. D.	12-31. A.	2-32. D.
12-33. C.	12-34. B.	12-35. B.	12-36. D.
12-37. D	12-38. D	12-39. B	

CHAPTER 13 : CHOLESTEROL, LIPOPROTEINS

13-1. All are useful substances produced from cholesterol, *except:*

A. Vitamin D B. Bile salts
C. Bile pigments D. Cortisol

13-2. The ring system of cholesterol is called:
A. Cholanthrene ring
B. Cyclopentanophenanthrene
C. Naphthoquinone
D. Corrin ring

13-3. During cholesterol biosynthesis, the first sterol ring synthesizes is:
A. Squalene
B. Zymosterol
C. Lanosterol
D. Desmosterol

13-4. The key enzyme in the pathway of cholesterol biosynthesis is:
A. HMG CoA synthase
B. HMG CoA lyase
C. HMG CoA reductase
D. Mevalonate kinase

13-5. Cholesterol contains how many carbon atoms?
A. 6 B. 12
C. 27 D. 30

13-6. The serum lipoprotein with the highest cholesterol content is:
A. Alpha lipoprotein (HDL)
B. Beta lipoprotein (LDL)
C. Prebeta lipoprotein (VLDL)
D. Chylomicrons

13-7. The main apoprotein present in the low density lipoprotein (LDL) is:
A. A and D B. B-100
C. E2 and C D. B-48

13-8. The main apoprotein present in chylomicron:
A. D B. B-100
C. E-2 D. B-48

13-9. The main apoprotein present in HDL is:
A. A-1 B. B-100
C. E-2 and C D. B-48

13-10. Function of chylomicron is the transport of:
A. Triglycerides from intestine to adipose tissue
B. Cholesterol from liver to peripheral tissues
C. Cholesterol from peripheral tissues to liver
D. Free fatty acids (NEFA) from adipose tissue

13-11. Function of HDL is the transport of:
A. Triglycerides from intestine to adipose tissue
B. Cholesterol from liver to peripheral tissues
C. Cholesterol from peripheral tissues to liver
D. Free fatty acids (NEFA) from adipose tissue

13-12. Function of LDL is the transport of:
A. Triglycerides from intestine to adipose tissue
B. Cholesterol from liver to peripheral tissues
C. Cholesterol from peripheral tissues to liver
D. Free fatty acids (NEFA) from adipose tissue

13-13. The cyclopentanophenanthrene ring structure is seen in all the following compounds, *except*:
A. Testosterone
B. Adrenocorticotropin
C. Glycocholic acid
D. Vitamin D

13-14. All are intermediates in the biosynthesis of cholesterol, *except*:
A. Acetyl CoA B. Cholyl CoA
C. Acetoacetyl CoA D. HMG CoA

13-15. All are true with respect to plasma lipoproteins, *except*:
A. Lipoprotein lipase is activated by apo-A-1
B. LCAT is inhibited by apo-B-100
C. Apo-E is involved in receptor mediated uptake of chylomicron remnants
D. Apo-A1 is atherogenic

13-16. Select the correct statement regarding HDL:
A. Normal HDL cholesterol in serum is 150–250 mg%
B. It contains apoprotein B-48
C. It is associated with enzymatic conversion of free cholesterol to cholesterol ester
D. Mature HDL has discoid shape

13-17. Function of albumin is the transport of:
A. Triglycerides from intestine to adipose tissue
B. Cholesterol from liver to peripheral tissues
C. Cholesterol from peripheral tissues to liver
D. Free fatty acids (NEFA) from adipose tissue to peripheral tissues

13-18. All are true in familial hyperlipoprotenemia Type II-A, *except*:
A. TAG level in blood is increased
B. LDL level in blood is increased
C. LDL receptor defect is seen
D. Cholesterol level in blood is increased

13-19. The plasma sample of a 35-year-old man after overnight refrigeration showed a creamy layer on top and opalescence below. The condition which is excluded is:
A. Diabetes mellitus
B. Alcoholism
C. Hypothryroidism
D. Familial hypercholesterolemia

Answers (Chapter 13)

13-1. C.	13-2. B.	13-3. C.	13-4. C.
13-5. C.	13-6. B.	13-7. B.	13-8. D.
13-9. A.	13-10. A.	13-11. C.	13-12. B.
13-13. B.	13-14. B.	13-15. C	13-16. C
13-17. D	13-18. A.	13-19. D.	

CHAPTER 14: MCFA, PUFA, PROSTAGLANDINS

14-1. Triglycerides containing medium chain fatty acids are hydrolyzed in the gastrointestinal tract by:
A. Epithelial triglyceride lipase
B. Pancreatic lipase
C. Hormone sensitive lipase
D. Lipoprotein lipase

14-2. Polyunsaturated fatty acids are required for the following purposes, *except*:
A. Synthesis of leukotrienes
B. Synthesis of prostaglandins
C. Esterification of cholesterol
D. Conjugation of bile acids

14-3. Desaturation of fatty acids is taking place in which subcellular fraction?
A. Mitochondria B. Cytoplasm
C. Endoplasmic reticulum D. Lysosome

14-4. All the statements regarding prostaglandin are correct, *except*:

A. Leukotrienes are derivatives of prostaglandins
B. Arachidonic acid is the precursor for PG2 series
C. Naturally occurring prostaglandins belong to PG2
D. Synthesis is inhibited by aspirin

14-5. The mechanism of anti-inflammatory action of aspirin is that it:

A. Reversibly inhibits cyclo-oxygenase
B. Irreversibly inhibits cyclo-oxygenase
C. Inhibits phospholipase
D. Inhibits lipo-oxygenase

14-6. Prostaglandins are stored:

A. In adipose tissue
B. As plasma lipoproteins
C. As granules in mast cells
D. As membrane phospholipids

14-7. Polyunsaturated fatty acids are required for the synthesis of the following compounds, *except*:

A. Lecithin B. Cholesterol
C. Prostacyclin D. Leukotrienes

14-8. Phosphatidylcholine is synthesized by combining diacetylglycerol with activated choline. Which derivative of choline is used for this reaction?

A. UDP-choline B. GDP-choline
C. ATP-choline D. CDP-choline

14-9. The lipid storage disease which can manifest in adults is:

A. Tay-Sach's disease
B. Niemann-Pick's disease
C. Sandhoff's disease
D. Gaucher's disease

Answers (Chapter 14)

14-1. A.	14-2. D.	14-3. C.	14-4. A.
14-5. B.	14-6. D.	14-7. B.	14-8. D.
14-9. D.			

CHAPTER 15: GENERAL METABOLISM OF AMINO ACIDS

15-1. Which enzyme is an exopeptidase?

A. Trypsin B. Chymotrypsin
C. Pepsin D. Carboxy peptidase

15-2. Trypsinogen is activated to trypsin by removal of a hexapeptide by:

A. Pepsin B. Chymotrypsin
C. Enterokinase D. Pancreozymin

15-3. All the following are transamination reactions, *except*:

A. Pyruvic acid to alanine
B. Oxaloacetic acid to aspartic acid
C. Hydroxy pyruvic acid to serine
D. Glutamic acid to glutamine

15-4. All are true with regard to transamination reactions, *except:*

A. It is the starting step of metabolic degradation of most of the amino acids
B. This process synthesize non-essential amino acids
C. It requires pyridoxal phosphate as co-enzyme
D. Reaction takes place only in liver

15-5. Ammonia is formed with the help of all the following enzymes, *except*:

A. Transaminases
B. Glutamate dehydrogenase
C. Amino acid oxidase
D. Threonine dehydratase

15-6. Which amino acid is oxidatively de-aminated in liver?

A. Aspartic acid B. Alanine
C. Glutamic acid D. Valine

15-7. During urea cycle, the two nitrogen atoms are derived from:

A. Ammonia and arginine
B. Ammonia and aspartic acid
C. Both from ammonia
D. Ammonia and ornithine

15-8. Ammonia is trapped in brain by:

A. Glutamine synthetase reaction
B. Glutaminase reaction
C. Urea synthesis cycle
D. Glutamate dehydrogenase reaction

15-9. Normal blood urea level is:

A. 10–20 mg/dL B. 20–40 mg/dL
C. 20–40 g/dL D. 30–60 mg/dL

15-10. Blood urea level is markedly increased in:

A. Liver diseases B. Renal diseases
C. Cardiac diseases D. Protein intake

15-11. The major donor of carbon atoms to the one-carbon pool is:

A. Serine B. Tyrosine
C. Threonine D. Proline

15-12. All are true with regard to transamination reactions, *except*:

A. It is the starting step of degradation of amino acids
B. This process synthesize non-essential amino acids
C. It requires pyridoxal phosphate as co-enzyme
D. Reaction takes place only in liver

15-13. All the following gastrointestinal enzymes are secreted as pro-enzymes (zymogens), *except*:

A. Trypsin B. Chymotrypsin
C. Pepsin D. Ribonuclease

15-14. Succinyl CoA is formed from following, *except:*

A. Valine B. Isoleucine
C. Propionyl CoA D. Aspartate

15-15. Pyridoxal phosphate is the co-enzyme of which reactions?

A. Transamination reactions
B. TGlutamate dehydrogenase
C. L-amino acid oxidase
D. Formimino glutamic acid to glutamate

15-16. All the amino acids contribute carbon atoms to the one-carbon pool, *except:*

A. Tryptophan B. Histidine
C. Valine D. Serine

Answers (Chapter 15)

15-1. D.	15-2. C.	15-3. D.	15-4. D.
15-5. A.	15-6. C.	15-7. B.	15-8. A.
15-9. B.	15-10. B.	15-11. A.	15-12. D.
15-13. D	15-14. D	15-15. A.	15-16. C.

CHAPTER 16: SIMPLE AMINO ACIDS

16-1. Glycine is used for synthesis of all the following compounds, *except:*
A. Serine B. Creatinine
C. Pyrimidine ring D. Heme

16-2. In the body, glycine is used for synthesis of all the following substances, *except:*
A. Purine ring B. Glutamine
C. Glutathione D. Creatine

16-3. Normal serum creatinine level is:
A. 0.2–0.4 mg/dL B. 0.3–0.6 mg/dL
C. 0.7–1.4 mg/dL D. 1.4–2.8 mg/dL

16-4. Creatine is synthesized from the following amino acids, *except*:
A. Arginine B. Aspartic acid
C. Glycine D. Methionine

16-5. The sources of oxalic acid in urine are:
A. Ornithine and citrulline
B. Oxalosuccinate and formic acid
C. Oxaloacetate and aspartic acid
D. Ascorbic acid and glycine

16-6. Serine when decarboxylated will produce:
A. Ethanolamine
B. Beta mecarptoethanolamine
C. Histamine
D. Putrescine

16-7. Serine is a precursor of all the following, *except:*
A. Methionine B. Glycine
C. Choline D. Cysteine

16-8. Acetylcholine is derived from which amino acid?
A. Tyrosine B. Tryptophan
C. Glutamic acid D. Serine

16-9. The methyl donor in methyl transfer reactions is:
A. Methyl cobalamin
B. Methyl malonyl CoA
C. S-adenosyl methionine
D. S-adenosyl homocysteine

16-10. All the following are substrates for transmethylation reactions, *except:*
A. Guanidinoacetic acid
B. Choline
C. Norepinephrine
D. N-acetyl serotonin

16-11. Name the defective enzyme in cystathioninuria:
A. Cystathionase
B. Phenylalanine hydroxylase
C. Homogentisic acid oxidase
D. Para-hydroxyphenylpyruvate oxidase

16-12. Cysteine on transamination gives rise to:
A. Pyruvic acid
B. Para-hydroxyphenylpyruvate
C. Phenyl pyruvate
D. Beta-mercaptopyruvate

16-13. Glutathione is used in the body for all the following functions, *except:*
A. For ammonia trapping
B. For reduction of methemoglobin to normal hemoglobin
C. To carry amino acids across membranes
D. For removal of peroxidases

16-14. One salient clinical manifestation in homocystinuria is:
A. Agitation, hyperkinesia
B. Renal stones
C. Ectopia lentis (subluxation of lens)
D. Nystagmus

16-15. The incidence of homocystinuria in the population is about:
A. 1 in 1,500
B. 1 in 50,000
C. 1 in 1,00,000
D. 1 in 2,00,000

16-16. Example for a transport defect is:
A. Cystinuria
B. Homocystinuria
C. Crigler-Najjar syndrome
D. Sulfituria

16-17. The urine of a patient with homocystinuria will be positive for:
A. Benedict's test
B. Ferric chloride test
C. Rothera's test
D. Cyanide nitroprusside test

16-18. Urine of a 12-year-old boy gave a positive cyanide nitroprusside test. He had renal stones. He is likely to have:
A. Homocystinuria B. Cystinosis
C. Hartnup's disease D. Renal glycosuria

16-19. Name defective enzyme in homocystinuria:
A. Propionyl CoA carboxylase
B. Methyl malonyl racemase
C. Tyrosinase
D. Cystathionine synthase

16-20. Following constituents of urine listed below contain Nitrogen derived from glycine, *except*:
A. Creatinine B. Uric acid
C. Indican D. Urobilinogen

16-21. Alanine on transamination gives rise to:
A. Pyruvic acid
B. Para-hydroxyphenylpyruvate
C. Phenyl pyruvate
D. Hydroxy pyruvate

16-22. S-adenosyl methionine transfers methyl groups to which compound?
A. Homocysteine B. Methionine
C. Homoserine D. Norepinephrine

16-23. All the following are excretory products of metabolism of methionine and cysteine, *except:*
A. Ethereal sulfates
B. Inorganic sulfates
C. Sulfites
D. Phosphates

Answers (Chapter 16)

16-1. C.	16-2. B.	16-3. C.	16-4. B.
16-5. D.	16-6. A.	16-7. A.	16-8. D.
16-9. C.	16-10. B.	16-11. A.	16-12. D.
16-13. A.	16-14. C.	16-15. D.	16-16. A.
16-17. D.	16-18. B.	16-19. D.	16-20. C.
16-21. A.	16-22. D.	16-23. D.	

CHAPTER 17: ACIDIC, BASIC AMINO ACIDS

17-1. Which amino acid will give rise to an inhibitory neurotransmitter?
A. Histidine B. Glutamic acid
C. Ornithine D. Tyrosine

17-2. Gamma amino butyric acid (GABA) is derived from which amino acid?
A. Tyrosine B. Tryptophan
C. Glutamic acid D. Serine

17-3. Transamination of glutamic acid leads to:
A. Glutamine
B. Alpha ketoglutaric acid
C. Beta hydroxy beta methyl glutaric acid
D. Gamma amino butyric acid

17-4. Transamination of aspartic acid gives rise to:
A. Asparagine B. Malate
C. Oxalosuccinate D. Oxaloacetate

17-5. Lysine is:
A. Deficient in cereals (rice, wheat)
B. Deficient in pulses (dal, Bengal gram)
C. Mainly glucogenic
D. A non-essential amino acid

17-6. Glutaminase which hydrolyzes glutamine to glutamic acid is used for:
A. Ammonia trapping in brain
B. Excretion of ammonium ions in renal tubules
C. Purine synthesis
D. Transamination

17-7. Which amino acid does not undergo transamination
A. Leucine B. Isoleucine
C. Lysine D. Alanine

17-8. Lysine is required for the synthesis of all the following, *except:*

A. Collagen
B. Cadaverine
C. Carnitine
D. Carnosine

17-9. Alpha amino group of aspartic acid is incorporated into all the following compounds, *except:*

A. Adenine
B. Cytosine
C. Urea
D. Creatine

17-10. Aspartic acid is used for the synthesis of which compound?

A. Porphyrin
B. Sphingomyelin
C. Pyrimidine
D. Folic acid

17-11. All are monoamines, *except:*

A. Histamine
B. Serotonin
C. Dopamine
D. Glutamine

17-12. Ornithine when decarboxylated will produce what?

A. Ethanolamine
B. Beta mecarptoethanolamine
C. Histamine
D. Putrescine

17-13. Branched chain keto acids are excreted in urine in large quantities in:

A. Phenylketonuria
B. Maple syrup urine disease
C. Tyrosinosis
D. Hartnup's disease

17-14. Which test will be positive with the urine of patient with maple syrup urine disease?

A. Dinitrophenylhydrazine test
B. Benedict's test
C. Ferric chloride test
D. Cyanide nitroprusside test

17.15. Name the defective enzyme in Maple syrup urine disease (branched chain ketonuria):

A. Cystathionase
B. Phenylalanine hydroxylase
C. Homogentisic acid oxidase
D. Oxidative decarboxylase

Answers (Chapter 17)

17-1. B.	17-2. C.	17-3. B.	17-4. D.
17-5. A.	17-6. B.	17-7. C.	17-8. D.
17-9. D.	17-10. C.	17-11. D.	17-12. D.
17-13. B.	17-14. A.	17-15. D.	

CHAPTER 18: AROMATIC AMINO ACIDS

18-1. Which amino acid will give rise to the major pigment of the body?
A. Histidine
B. Glutamic acid
C. Ornithine
D. Tyrosine

18-2. Epinephrine is derived from which amino acid?
A. Tyrosine
B. Tryptophan
C. Glutamic acid
D. Serine

18-3. In pheochromocytoma, there is the excessive urinary excretion of:
A. Vanillylmandelic acid
B. Polyamines
C. Hippuric acid
D. 5-hydroxyindoleacetic acid

18-4. Name the defective enzyme in phenylketonuria:
A. Cystathionase
B. Phenylalanine hydroxylase
C. Homogentisic acid oxidase
D. Para-hydroxyphenylpyruvate oxidase

18-5. One salient clinical manifestation in phenylketonuria is:
A. Agitation, hyperkinesia
B. Renal stones
C. Ectopia lentis (subluxation of lens)
D. Nystagmus

18-6. The incidence of phenylketonuria in the population is about:
A. 1 in 1,500
B. 1 in 50,000
C. 1 in 1,00,000
D. 1 in 2,00,000

18-7. Albinism is due to the deficiency of the enzyme:
A. Phenylalanine hydroxylase
B. Tyrosinase
C. Para-hydroxyphenylpyruvate oxidase
D. Tyrosine dehydrogenase

18-8. All the following inborn errors of metabolism are characterized by mental retardation, *except:*

A. Phenyl ketonuria
B. von-Gierke's disease
C. Alkaptonuria
D. Maple syrup urine disease

18-9. Name the defective enzyme in alkaptonuria:

A. Cystathionase
B. Phenylalanine hydroxylase
C. Homogentisic acid oxidase
D. Para-hydroxyphenylpyruvate oxidase

18-10. All the following are synthesized from tyrosine, *except:*

A. Melanin
B. Tri-iodothyronine
C. Norepinephrine
D. Tryptamine

18-11. Homogentisic acid is excreted in urine in:

A. Phenylketonuria
B. Maple syrup urine disease
C. Tyrosinosis
D. Alkaptonuria

18-12. Xanthurenic acid excretion is seen in:

A. Pyridoxine deficiency
B. Hartnup disease
C. Gout
D. Xanthine oxidase deficiency

18-13. Which amino acid will give rise to a vitamin?

A. Tyrosine
B. Tryptophan
C. Glutamic acid
D. Histidine

18-14. Serotonin is derived from which amino acid?

A. Tyrosine
B. Tryptophan
C. Glutamic acid
D. Histidine

18.15. Melatonin is synthesized from:

A. Tyrosine
B. Tryptophan
C. Glutamic acid
D. Histidine

18-16. Pellagra is a manifestation of all the following conditions, *except:*

A. Thiamine deficiency
B. Pyridoxal deficiency
C. Carcinoid syndrome
D. Tryptophan deficiency

18-17. Tryptophan gives rise to all the following, *except:*
A. Alanine B. Melanin
C. Niacin D. Serotonin

18-18. All the following amino acids are both glucogenic and ketogenic, *except:*
A. Leucine B. Isoleucine
C. Tyrosine D. Tryptophan

18-19. Alpha ketoglutaric acid may be formed from all the following, *except:*
A. Glutamic acid B. Histidine
C. Proline D. Succinyl CoA

18-20. Name the defective enzyme in alkaptonuria:
A. Cystathionase
B. Phenylalanine hydroxylase
C. Homogentisic acid oxidase
D. Para-hydroxyphenylpyruvate oxidase

18-21. Ochronosis is seen in:
A. Phenylketonuria B. Alkaptonuria
C. Tyrosinosis D. Albinism

18-22. Tyrosine on transamination gives rise to:
A. Pyruvic acid
B. Para-hydroxyphenylpyruvate
C. Phenyl pyruvate
D. Hydroxy pyruvate

18-23. Xanthurenic acid in urine is seen in all conditions, *except:*
A. Pyridoxal deficiency
B. Niacin deficiency
C. Tryptophan deficiency
D. Tyrosine deficiency

18-24. Which of the following is synthesized from tryptophan?
A. Serotonin B. Melanin
C. Epinephrine D. Thyroxin

18-25. Tetrahydrofolic acid is the co-enzyme of:
A. Transamination reactions
B. Glutamate dehydrogenase
C. L-amino acid oxidase
D. Formimino glutamic acid to glutamic acid

18-26. Histidine when decarboxylated will produce:
A. Ethanolamine
B. Beta mecarptoethanolamine
C. Histamine
D. Putrescine

Answers (Chapter 18)

18-1. D.	18-2. A.	18-3. A.	18-4. B.
18-5. A.	18-6. A.	18-7. B.	18-8. C.
18-9. C.	18-10. D.	18-11. D.	18-12. A.
18-13. B	18-14. B.	18-15. B.	18-16. A.
18-17. B.	18-18. A.	18-19. D.	18-20. C.
18-21. B.	18-22. B.	18-23. D.	18-24. A.
18-25. D.	18-26. C.		

CHAPTER 19: CITRIC ACID CYCLE

19-1. All the enzymes catalyze reactions involving removal of carbon dioxide, *except*:
A. Alpha ketoglutarate dehydrogenase
B. Glutamate dehydrogenase
C. Pyruvate dehydrogenase
D. Isocitrate dehydrogenase

19-2. Which intermediate of TCA cycle is an unsaturated dicarboxylic acid?
A. Succinate
B. Malate
C. Oxaloacetate
D. Fumarate

19-3. Glucose has 6 carbon atoms; these are removed as carbon dioxide in all the following steps, *except:*
A. Succinate dehydrogenase reaction
B. Pyruvate dehydrogenase reaction
C. Isocitrate dehydrogenase step
D. Alpha ketoglutarate dehyrogenase step

19-4. How many ATPs are generated per one rotation of the citric acid cycle? (how many ATPs are produced when one molecule of acetyl CoA is oxidized in TCA cycle?)
A. 2
B. 8
C. 10
D. 15

19-5. All the following reactions involve in the generation of NADH, *except:*

A. Isocitrate to alpha ketoglutarate
B. Alpha ketoglutarate to succinyl CoA
C. Fumarate to malate
D. Malate to oxaloacetate

19-6. Acetyl CoA is used for the production of all the following, *except:*

A. Oxidation in TCA cycle
B. De novo synthesis of fatty acid
C. Glucose
D. Cholesterol

19-7. Acetyl CoA is produced from all the following, *except:*

A. Pyruvate
B. Beta oxidation of fatty acids
C. Leucine
D. Aspartic acid

19-8. Succinyl CoA is formed from all the following, *except:*

A. Aspartic acid B. Isoleucine
C. Valine D. Odd chain fatty acids

19-9. Fumarate is produced from all the following, *except:*

A. Malate B. Arginino succinate
C. Phenylalanine D. Tryptophan

19-10. TCA cycle is regulated by the availability of:

A. ATP
B. Glucose
C. Fatty acid
D. Succinate dehydrogenase

19-11. Which is the substrate level phosphorylation step in the TCA cycle?

A. Alpha ketoglutarate to succinyl CoA
B. Succinyl CoA to succinate
C. Succinate to fumarate
D. Malate to oxaloacetate

19-12. During starvation, all the following substances are increased in blood or tissues, *except:*

A. Ketone bodies B. Epinephrine
C. Glycogen D. Glucagon

19-13. Which of the following enzymes is confined to the mitochondria?
A. Lactate dehydrogenase (LDH)
B. Malate dehydrogenase (MDH)
C. Glucose-6-phosphate dehydrogenase
D Succinate dehydrogenase (SDH)

19-14. Starvation activates all the enzymes, *except:*
A. Alanine amino transferase
B. Carnitine acyl transferase
C. HMG CoA lyase
D. ATP citrate lyase

Answers (Chapter 19)

19-1. B.	19-2. D.	19-3. A.	19-4. C.
19-5. C.	19-6. C.	19-7. D.	19-8. A.
19-9. D.	19-10. A.	19-11. B.	19-12. C.
19-13. D.	19-14. D		

CHAPTER 20: ELECTRON TRANSPORT CHAIN

20-1. All contain high energy bond, *except:*
A. ATP
B. Glucose-6-phospahte
C. Acetyl CoA
D. Phosphoenol pyruvate

20-2. All the following phosphate esters are high energy compounds, *except:*
A. Creatine phosphate
B. Carbamoyl phosphate
C. 2,3-bisphosphoglycerate
D. 1,3-bisphosphoglycerate

20-3. Co-enzyme Q catalyzes electron transport between:
A. FADH and cytochrome b
B. It is the last member in the electron transport chain
C. NADH and ubiquinone
D. Cytochrome Q and cytochrome C

20-4. Which contains copper?
A. Cytochrome oxidase
B. Cytochrome b5
C. Co-enzyme Q
D. Cytochrome P-450

20-5. Death due to cyanide poisoning is a result of:

A. Cyanide hemoglobin complex formation
B. Cyanide inhibiting complex I of respiratory chain
C. Cyanide inhibiting cytochrome oxidase
D. Cyanide blocking oxygen transport in blood

20-6. All are true with ATP synthase, *except:*

A. It has two subunits, F1 and F0
B. F1 subunit has catalytic activity
C. F0 subunit serves as a proton channel
D. Proton flow is from F1 to F0

20-7. Oxidative phosphorylation is inhibited by the following, *except:*

A. Oligomycin B. Carbon monoxide
C. Hydrogen cyanide D. Pyrophosphate

20-8. Which is the inborn error due to a mutation in NADH-Q reductase?

A. Amyotrophic lateral sclerosis
B. Leber's optic myopathy
C. Ragged red fiber disease
D. Duchenne muscular dystrophy

20-9. Valinomycin inhibits oxidative phosphorylation, because:

A. It inhibits ATP synthase
B. It inhibits cytochrome oxidase
C. It forms a complex with NADH
D. It makes mitochondria permeable to potassium

20-10. Which is true with regard to oligomycin?

A. It inhibits oxidative phosphorylation
B. It is an uncoupler
C. It inhibits translocase
D. It acts as an ionophore

20-11. Which is true with cytochrome reductase?

A. It catalyzes the electron transport between FADH and cytochrome b
B. It is the last member in the electron transport chain
C. It catalyzes the electron transport between NADH and ubiquinone
D. It catalyzes the electron transport between Co Q and cytochrome C

20-12. Which is an NAD^+ linked dehydrogenase?

A. Succinate dehydrogenase
B. Fatty acyl CoA dehydrogenase
C. Malate dehydrogenase
D. Glucose-6-phosphate dehydrogenase

20-13. Which of the following compound does not contain a high energy bond?

A. Fructose-1,6-bisphosphate
B. 1,3-bisphosphoglycerate
C. Succinyl CoA
D. Creatine phosphate

20-14. All are true with regard to mitochondria, *except*:

A. Glycolytic enzymes are in mitochondrial matrix
B. Inner mitochondrial membrane contains enzymes of electron transport chain
C. Fluid matrix contains enzymes of TCA cycle
D. Acyl CoA synthetase is localized in the outer mitochondrial membrane

Answers (Chapter 20)

20-1. B.	20-2. C.	20-3. A.	20-4. A.
20-5. C.	20-6. D.	20-7. D.	20-8. B.
20-9. D.	20-10. A.	20-11. D.	20-12. C.
20-13. A.	20-14. A.		

CHAPTER 21: HEME SYNTHESIS AND BREAKDOWN

21-1. All the following oxygen handling proteins contain heme as a prosthetic group, *except:*

A. Cytochromes
B. Peroxidase
C. Superoxide dismutase
D. Tryptophan pyrrolase

21-2. The methenyl bridge of protoporphyrin is derived from:

A. Alpha carbon atom of glycine
B. Alpha carbon atom of succinyl CoA
C. Carboxyl group of succinyl CoA
D. Carboxyl group of glycine

21-3. Rate limiting enzyme in heme synthesis is:
A. Heme synthase
B. ALA dehydrase
C. Uroporphyrinogen synthase
D. ALA synthase

21-4. All the following are precursors of heme, *except:*
A. Porphobilinogen B. Urobilinogen
C. Uroporphyrinogen D. Protoporphyrinogen

21-5. Which enzyme of heme biosynthesis is inhibited by lead?
A. ALA dehydratase B. UBG I synthase
C. UBG decarboxylase D. PPG oxidase

21-6. Acute intermittent porphyria (AIP) is:
A. Characterized by photodermatitis
B. An autosomal dominant trait
C. More common in men than women
D. Hemolytic anemia is associated

21-7. Congenital erythropoietic porphyria is caused by deficiency of:
A. ALA synthase
B. Uroporphyrinogen III co-synthase
C. Co-proporphyrinogen oxidase
D. Uroporphyrinogen decarboxylase

21-8. Urine of a patient with acute intermittent porphyria is likely to contain:
A. Porphobilinogen B. Uroporphyrinogen
C. Protoporphyrinogen D. Bilirubin

21-9. Degradation of heme to bilirubin releases one molecule of:
A. Carbon dioxide B. Water
C. Ammonia D. Carbon monoxide

21-10. Heme is converted to bilirubin in:
A. Mitochondria B. Microsomes
C. Golgi bodies D. Plasma membrane

21-11. Bilirubin in serum can be measured by:
A. Van den Bergh reaction
B. Ehrlich's reaction
C. Schlesinger's reaction
D. Fouchet's reaction

21-12. Which disease is associated with conjugated hyperbilirubinemia?

A. Gilbert's disease
B. Crigler-Najjar syndrome
C. Dubin-Johnson's syndrome
D. Congenital spherocytosis

21-13. All the following have a regulatory effect on heme synthesis, *except:*

A. Cellular heme pool
B. Low calorie diet
C. Compartmentalization of enzymes
D. Hematocrit value

21-14. Which intermediate of heme synthetic pathway is not a tetrapyrole?

A. Uroporphyrinogen
B. Coproporphyrinogen
C. Porphobilinogen
D. Protoporphyrinogen

21-15. Bilirubin in serum can be measured by:

A. Van den Bergh reaction
B. Ehrlich's reaction
C. Schlesinger's reaction
D. Fouchet's reaction

21-16. Porphyrins are:

A. Monopyrroles
B. Colorless components
C. Showing fluorescence under UV light
D. Destroyed by sunlight

21-17. In obstructive jaundice, prothrombin time:

A. Remains normal
B. Decreases
C. Becomes normal when vitamin K is administered
D. Increases when vitamin K is administered

21-18. Which does not contain heme?

A. Myoglobin B. Cytochrome
C. Ferroxidase D. Peroxidase

21-19. All are correct with regard to heme, *except:*

A. Ferroprotoporphyrin
B. Prosthetic group of cytochromes
C. Synthesized by using glycine
D. Linear tetrapyrrole

21-20. Which is required for synthesis of porphyrins?
A. Ammonia B. Methionine
C. Glycine D. Carbon dioxide

21-21. All are substituent groups present in heme (protoporphyrin), *except:*
A. Methyl B. Vinyl
C. Propionyl D. Ethyl

21-22. All are correct with heme synthesis, *except:*
A. Compartmentalized
B. Stimulated by lead
C. Regulated by end product inhibition
D. Induced by barbiturates

21-23. All affect heme synthesis, *except:*
A. Barbiturates B. Estrogens
C. Glucose D. Oncogenes

21-24. A patient with acute intermittent porphyria should be given:
A. High carbohydrate diet
B. Barbiturates
C. Aspirin
D. Saline infusion

21-25. Congenital erythropoietic porphyria has all the features, *except:*
A. Sensitivity to sunlight
B. Reddish discoloration of urine
C. Erythrocyte porphyrin level elevated
D. Red fluorescence of teeth

21-26. Presence of porphyrins in urine can be detected by:
A. UV fluorescence
B. Absorption band at 565 nm
C. Fouchet's test
D. Schlesinger's test

21-27. Bilirubin is formed in all the tissues, *except:*
A. Liver B. Spleen
C. Kidney D. Bone marrow

21-28. Which is not involved in catabolism of heme?
A. Cytochrome a B. NADPH
C. Molecular oxygen D. Cytochrome c

21-29. Which enzyme is not involved in catabolism of heme?

A. Heme oxygenase
B. UDP-glucuronyl transferase
C. Biliverdin reductase
D. Ferrochelatase

21-30. All the following are produced from heme during catabolism, *except*:

A. Urobilinogen B. Porphobilinogen
C. Bilirubin D. Biliverdin

21-31. Normal color of feces is due to:

A. Bilirubin B. Stercobilin
C. Porphobilinogen D. Biliverdin

21-32. All the following compounds undergo enterohepatic circulation, *except*:

A. Urobilinogen B. Iron
C. Cholic acid D. Cholesterol

21-33. Increased level of urobilinogen in urine is observed in all the following conditions, *except:*

A. Congenital spherocytosis
B. Mismatched transfusion
C. Glucose-6-phosphate dehydrogenase deficiency
D. Lead poisoning

21-34. Congenital hyperbilirubinemia with a bilirubin level of 2.5 mg/dL is suggestive of:

A. Criggler-Najjar syndrome
B. Gilbert's syndrome
C. Dubin-Jhonson's syndrome
D. Rotor syndrome

Answers (Chapter 21)

21-1. C.	21-2. A.	21-3. D.	21-4. B.
21-5. A.	21-6. B.	21-7. B.	21-8. A.
21-9. D.	21-10. B.	21-11. A.	21-12. C.
21-13. D.	21-14. C.	21-15. A.	21-16. C.
21-17. C.	21-18. C	21-19. D.	21-20. C.
21-21. D.	21-22. B.	21-23. D.	21-24. A.
21-25. C.	21-26. A.	21-27. C.	21-28. A.
21-29. D.	21-30. B.	21-31. B.	21-32. B.
21-33. D.	21-34. B.		

CHAPTER 22: HEMOGLOBIN

22-1. HbA differs from HbF in that:
A. HbA has only alpha chains
B. HbF cannot bind to 2,3-BPG
C. HbF can bind only 2 molecules of oxygen
D. HbA is alkali resistant

22-2. Methemoglobinemia is found in all the following conditions, *except:*
A. Ingestion of nitrites
B. Carbon monoxide poisoning
C. Presence of Hb M
D. Poisoning by aniline dyes

22-3. All the following are true with regard to carboxyhemoglobin, *except:*
A. CO binds to ferrous iron to form carboxy-hemoglobin
B. CO has more affinity for Hb than oxygen
C. Oxygen dissociation curve is shifted to right
D. Cannot be converted back to deoxy Hb

22-4. One deciliter (100 mL) of blood with 15 grams of hemoglobin can carry how much oxygen?
A. 7.5 mL of oxygen
B. 15 mL of oxygen
C. 20 mL of oxygen
D. 30 mL of oxygen

22-5. Which factor causes a shift in the ODC (oxygen dissociation curve) to left?
A. High pO_2
B. Low pH
C. Low pO_2
D. High temperature

22-6. All the following manifestations are seen in sickle cell anemia (HbS disease), *except:*
A. Pain and swellings in joints
B. Hemolytic anemia
C. Sickled cells in peripheral circulation
D. Inclusion bodies in RBCs

22-7. A person with HbS trait is likely to have:
A. About 90% of hemoglobin as HbS
B. A single band on electrophoresis
C. Sickling at high altitudes
D. Irreversibly sickled cells in peripheral smear

22-8. All are features of isohydric transport, *except*:
A. CO_2 is transported in plasma as bicarbonate
B. Carbonic anhydrase is the major enzyme involved
C. Chloride shift is a result of CO_2 transport
D. CO_2 is bound to amino terminal group

22-9. Methemoglobinemia may be caused by all the following, *except*:
A. Methemoglobin reductase deficiency
B. Presence of Hb M
C. Ingestion of nitrites
D. Deficiency of ascorbic acid

22-10. Carbon dioxide is carried in blood with all the forms, *except*:
A. Dissolved form
B. Isohydric transport
C. Carboxyhemoglobin
D. Carbaminohemoglobin

22-11. All are true with regard to HbF, *except*:
A. Poorly binds 2,3-BPG
B. Co-polymerizes with HbS
C. Affinity for oxygen is more
D. Is a tetramer of 2 alpha and 2 gamma chains

22-12. Regarding myoglobin, all are true, *except*:
A. It has 4 polypeptide chains
B. Binds oxygen
C. Oxygen binding affinity is higher than that of hemoglobin
D. Contains one heme residue per molecul

22-13. Regarding HbS and HbA, all are true, *except*:
A. The chain length of HbS is smaller than HbA
B. Solubility of deoxy HbS is lower than deoxy HbA
C. HbS is slower moving on electrophoresis than HbA
D. Structural difference is with regard to beta chain

22-14. Anemia can result from deficiency of all the following, *except*:
A. Iron
B. Cobalamin
C Cholecalciferol
D. Pteroyl glutamic acid

22-15. The common molecular defects in beta thalassemia include the following, *except:*

A. Frame shift mutations
B. Nonsense mutations
C. Suppressor mutations
D. Improper splicing of hnRNA

22-16. Unstable hemoglobins will produce:

A. Erythrocytosis B. Heinz body anemia
C. Polycythemia D. Sickling

22-17. Presence of an abnormal hemoglobin is suspected if a patient has any of the signs, *except:*

A. Nonspherocytic hemolytic anemia
B. Hypochromic microcytic anemia
C. Polycythemia
D. Cyanosis without cardiopulmonary dysfunction

22-18. Hemoglobin will not co-polymerize with HbS?

A. Hb C B. Hb A
C. Hb D D. Hb E

22-19. Which phenomenon is not related to functional properties of hemoglobin?

A. Bohr effect B. Haldane effect
C. Pasteur effect D. Donnan effect

22-20. In homozygous beta thalassemia:

A. Hb H is the major hemoglobin
B. Symptoms appear at birth
C. Persistence of Hb F is observed
D. Patients have a normal lifespan

22-21. Beta thalassemia can result from, *except:*

A. Terminator codon mutations
B. Abnormal splicing of mRNA
C. Defect in post-translational processing
D. Mutation in suppressor genes

22-22. Myoglobinuria is observed in all the following conditions, *except:*

A. Severe muscular exercise
B. Crush injuries of limbs
C. Myasthenia gravis
D. Muscle wasting diseases

22-23. Which hemoglobin derivative is normally found in blood?

A. Deoxyhemoglobin B. Sulfhemoglobin
C. Acid hematin D. Cyanmethemoglobin

22-24. Methemoglobinemia may be caused by all the following, *except:*

A. Methemoglobin reductase deficiency
B. Presence of Hb M
C. Ingestion of nitrites
D. Deficiency of ascorbic acid

22-25. All are the characteristics of Hb S, *except* it:

A. Is an abnormal hemoglobin
B. Is less soluble in deoxy state
C. Tends to polymerize when pO_2 is low
D. Cannot bind oxygen

22-26. A sudden shift in oxygen dissociation curve to the left can occur due to:

A. Diabetes mellitus
B. Acute renal failure
C. Metabolic alkalosis
D. Ascending to high altitudes

22-27. All are characteristics of thalassemia major, *except*:

A. It is manifested when the affected genes are in homozygous state
B. It is the result of gene deletion
C. It is a common molecular disease in India
D. It is incompatible with life

22-28. Which factor decreases the affinity of hemoglobin for oxygen binding?

A. An increase in 2,3-BPG level in blood
B. A rise in pH
C. Oxidation of iron to ferric state in one subunit
D. Vomiting

22-29. A patient with heterozygous alpha thalassemia will have:

A. Anemia
B. Hb H
C. Hb F
D. Hematuria

22-30. Patients having hemoglobin variants with increased oxygen affinity will have:

A. Increased level of Hb F
B. Shifting of ODC (oxygen dissociation curve) to left
C. Target cells in circulation
D. Binds 2,3-BPG with more affinity

22-31. All are correct with Hb S trait, *except:*

A. It confers a biological advantage against malaria

B. It is compatible with life

C. It is manifested at high altitudes as vaso-occlusive phenomenon

D. Chronic hemolytic anemia is the usual manifestation

22-32. All are true with regard to hemoglobin M variants, *except:*

A. They have increased oxygen carrying capacity

B. Are oxidized to methemoglobin

C. Binds only 2 molecules of oxygen

D. Responds to administration of methylene blue

22-33. During oxygen transport by hemoglobin:

A. One hemoglobin molecule carries 4 oxygen atoms

B. The graph of percentage of oxygen saturation versus oxygen pressure is sigmoid in shape

C. Oxygen binding causes the iron to change from ferrous to ferric state

D. Hemoglobin carries 1.34 mL of oxygen per 100 mL of blood

22-34. Regarding the CO_2 transport in blood:

A. OxyHb is more negatively charged (stronger acid) than deoxy Hb

B. In normal blood, bicarbonate to carbonic acid ratio is 1:20

C. One millimole deoxy hemoglobin can take up 10 mEq of hydrogen ions

D. Carboxyl end group of Hb chain takes up carbon dioxide

22.35. In HbA, the 6th amino acid is glutamic acid. In HbS, this is replaced by valine. The mobility of HbA and HbS during electrophoresis will be

A. HbS and HbA have same mobility

B. HbS moves faster than HbA to anode

C. HbS moves faster than HbA to cathode

D. HbS moves slower than HbA to anode

22-36. A person with sickle cell trait can follow any of the following occupations, *except:*
A. Software development
B. Biotechnologist
C. Fighter pilot
D. Air traffic controller

22-37. Toxic methemoglobinemia is characterized by all, *except:*
A. Severe central cyanosis
B. Cherry red color of blood
C. Blood gives a strong absorption band at 630 nm
D. Cyanosis improves on giving methylene blue

Answers (Chapter 22)

22-1. B.	22-2. B.	22-3. C.	22-4. C.
22-5. C.	22-6. D.	22-7. C.	22-8. D.
22-9. D.	22-10. C.	22-11. B.	22-12. A.
22-13. A.	22-14. C.	22-15. C.	22-16. B.
22-17. B.	22-18. B.	22-19. C.	22-20. C.
22-21. D.	22-22. C.	22-23. A.	22-24. D.
22-25. D.	22-26. C.	22-27. D.	22-28. A.
22-29. A.	22-30. B.	22-31. D.	22-32. A.
22-33. B.	22-34. A.	22-35. D.	22.36. C.
22-37. C.			

CHAPTER 23: CLINICAL ENZYMOLOGY

23-1. In myocardial infarction:
A. Serum LDH-1 level is lower than LDH-2
B. Increase in serum ALT level is diagnostic
C. Serum CK-MB levels show a marked increase
D. Serum alkaline phosphatase (ALP) is increased

23-2. How many iso-enzyme forms are there for lactate dehydrogenase (LDH)?
A. Two B. Three
C. Four D. Five

23-3. Serum alkaline phosphatase is increased in:
A. Hypothyroidism B. Carcinoma of prostate
C. Bone diseases D. Myocardial infarction

23-4. ALP (alkaline phosphatase) level in serum is elevated in all the following conditions, *except:*
A. Obstructive jaundice
B. Rickets
C. Infective hepatitis
D. Myocardial infarction

23-5. Estimation of gamma glutamyl transferase is useful to detect one of the following clinical conditions:
A. Alcohol abuse
B. Infective hepatitis
C. Kidney failure
D. Myocardial infarction

23-6. Serum acid phosphatase (ACP) level is increased in:
A. Rickets
B. Hepatoma
C. Prostatic carcinoma
D. Excess phosphate intake

23-7. Acute pancreatitisis diagnosed by estimating the blood concentration of which enzyme:
A. Alkaline phosphatase
B. Acid phosphatase
C. Alanine transaminase
D. Amylase

23-8. Iso-enzymes are:
A. Examples of polymorphism exhibited by the enzymes
B. Having the same amino acid sequence
C. Physically different but have identical catalytic properties
D. Polypeptide chains that differ by a single amino acid replacement

23-9. Creatine kinase level in serum is increased in:
A. Myocardial infarction
B. Infective hepatitis
C. Prostate cancer
D. Intravascular hemolysis

23-10. All are true with regard to Regan iso-enzyme of alkaline phosphatase, *except:*
A. It moves along with alpha-2 band in electrophoresis
B. Is heat stable (not destroyed at 65°C for 30 min)
C. Not inhibited by fluoride
D. Seen in carcinoma of lung

23-11. Which one of the following is functional enzyme in plasma?
A. Amylase
B. Lipase
C. Alkaline phosphatase
D. Plasminogen

23-12. Iso-enzymes may be identified or separated by all the following characteristics, *except:*
A. Electrophoretic mobility
B. Heat stability
C. Difference in Km value
D. Chemical reaction catalyzed

23-13. Enzymes may be used for following, *except:*
A. As therapeutic agents
B. For diagnosis
C. As nutrients
D. As tumor markers

23-14. Serum enzymes helpful in the diagnosis of myocardial infarction are the following, *except:*
A. CK-MB
B. Troponin
C. Glucose-6-phosphate dehydrogenase
D. LDH

23-15. All the following statements are true with regard to liver diseases, *except:*
A. In obstructive jaundice, alkaline phosphatase level is markedly increased
B. In infective hepatitis, alanine transaminase level is increased
C. In alcohol abuse, gamma glutamyl transpeptidase level is markedly increased
D. In cirrhosis, creatine phosphokinase level is increased

23-16. Pseudocholinesterase deficiency is manifested

A. Prolonged apnea after succinylcholine administration
B. RBC lifespan is reduced
C. Myasthenia gravis
D. Peripheral neuropathy

23-17. Serum acid phosphatase (ACP) level is increased in all the following conditions, *except:*

A. Metastasis in bones from carcinoma of prostate
B. Bone metastasis from carcinoma of breast
C. Per rectal examination of patient
D. Hemolyzed blood samples

Answers (Chapter 23)

23-1. C.	23-2. D.	23-3. C.	23-4. D.
23-5. A.	23-6. C.	23-7. D.	23-8. C.
23-9. A.	23-10. C.	23-11. D.	23-12. D.
23-13. C.	23-14. C.	23-15. D.	23-16. A.
23-17. B.			

CHAPTER 24: REGULATION OF BLOOD SUGAR

24-1. The normal fasting plasma glucose level is:

A. 40–6 mg/100 mL
B. 70–11 mg/100 mL
C. 120–15 mg/100 mL
D. 160–18 mg/100 mL.

24-2. Insulin activates:

A. Lipolysis B. Ketogenesis
C. Gluconeogenesis D. Glycolysis

24-3. When a standard oral glucose tolerance test is done, the blood glucose levels of the patient were found to be as follows: 0 min (fasting) = 120 mg/dL; 60 min = 170 mg/dL; 120 min = 150 mg/dL. The patient has:

A. Normal glucose tolerance
B. Impaired glucose tolerance
C. Mild diabetes mellitus
D. Severe diabetes mellitus

24-4. A comatose patient with tremors admitted to the hospital has a blood glucose value of 300 mg/dL and his urine was positive for both Rothera's test and Benedict's test. He is suffering from:
A. Untreated diabetes mellitus with ketosis
B. Diabetes mellitus treated with overdose of insulin
C. Renal glycosuria
D. Thyrotoxicosis

24-5. Activities of all the following enzymes are enhanced in starvation, *except:*
A. Phosphoenol pyruvate carboxykinase
B. Carnitine acyl transferase
C. Pyruvate carboxylase
D. Acetyl CoA carboxylase

24-6. Insulin increases activity of all the following enzymes, *except:*
A. Acetyl CoA carboxylase
B. Hormone sensitive lipase
C. Glycogen synthase
D. Glucose-6-phosphate dehydrogenase

24-7. The following are lipolytic hormones, *except:*
A. Glucagon
B. Cortisol
C. Epinephrine
D. Insulin

24-8. Growth hormone causes:
A. Decreased peripheral utilization of glucose
B. Decreased hepatic production of glucose by gluconeogenesis
C. Increased glycolysis in muscle
D. Decreased lipolysis

24-9. All are true with regard to Glucagon, *except:*
A. Increases lipolysis
B. Stimulates glycogenolysis
C. Favors gluconeogenesis
D. Produces glycogen synthesis

24-10. All are true with regard to glycohemoglobin, *except:*
A. Seen as high levels in sickle cell anemia
B. Otherwise known as Hb A1c
C. Useful index of long-term control of blood glucose level
D. About 3–5 of hemoglobin in normal cases

24-11. A lactating mother's urine was found to be positive for Benedict's test. Her fasting blood sugar was 80 mg/dL and post-prandial blood sugar after one and half hours after a breakfast was 140 mg/dL. She may be a case of:
A. Untreated diabetes mellitus
B. Diabetes mellitus treated with inadequate dose of insulin
C. Lactosuria
D. Renal glucosuria

24-12. Diabetes mellitus is caused by:
A. Anti-insulin antibodies
B. Low levels of glucocorticoids
C. Receptor abnormalities on target tissues
D. Reduced secretion of growth hormone

24-13. Which tissue is most insulin sensitive?
A. Brain
B. Liver
C. Adipose tissue
D. Cardiac muscle

24-14. Which laboratory finding is abnormal in a pregnant woman at 28 weeks gestation
A. Serum alkaline phosphatase level of 25 KAU/dL
B. Plasma 1hr post-prandial glucose of 200 mg/dL
C. Alpha fetoprotein level of 120 ng/mL
D. Plasma triglyceride level of 150 mg/dL

24-15. In the enzymatic method of glucose estimation, the following reactants are used, *except*:
A. Glucose oxidase
B. NAD^+
C. Peroxidase
D. Hydrogen peroxide

24-16. An apparently healthy man, on a routine check up, was found to have fasting blood sugar of 80 mg/dL, and urine showed no abnormal constituents. One and half hours after a heavy breakfast, his blood sugar was 120 mg/dL and urine sample at that time was positive for Benedict's test. He is suffering from:

A. Untreated diabetes mellitus
B. Diabetes mellitus treated with inadequate dose of insulin
C. Diabetes mellitus treated with overdose of insulin
D. Renal glucosuria

24-17. Benedict's reagent will be reduced by all the following substances seen in urine, *except:*

A. Fructose
B. Galactose
C. Pentoses
D. Sucrose

24-18. All affect glucose tolerance test, *except:*

A. Liver diseases
B. Acute infections
C. Vitamin B_{12} deficiency
D. Hyperthyroidism

24.19. Which statement is true with regard to insulin?

A. Secretion is from the beta cells of pancreas
B. Insulin favors gluconeogenesis
C. Insulin molecule has two chains connected by disulfide bridges
D. Insulin level in blood is regulated by the growth hormone level

24.20. All the following cellular activities are decreased under the influence of insulin, *except:*

A. Protein breakdown
B. Ketogenesis
C. Glycogen synthesis
D. Lipolysis

24-21. Insulin increases activity of all the following enzymes, *except:*

A. Acetyl CoA carboxylase
B. Hormone sensitive lipase
C. Glycogen synthase
D. Glucose-6-phosphate dehydrogenase

24-22. Insulin secretion is mainly controlled by:
A. Blood lipid level
B. Hypothalamic mechanism
C. HCl in stomach
D. Glucose level in blood

24-23. All are true with Insulin receptor, *except:*
A. It is a glycoprotein
B. Alpha subunit has tyrosine kinase activity
C. It has 4 subunits
D. Alpha subunits are located on the exterior of cell membrane

24-24. The following are lipolytic hormones, *except:*
A. Glucagon
B. Cortisol
C. Epinephrine
D. Insulin

24.25. When glucagon binds to its receptors on the liver cells, the following changes take place, *except:*
A. Activation of adenylate cyclase
B. Increase in concentration of cAMP
C. Inactivation of glycogen phosphorylase
D. Activation of protein kinase

24-26. Growth hormone causes:
A. Decreased peripheral utilization of glucose
B. Decreased hepatic production of glucose by gluconeogenesis
C. Increased glycolysis in muscle
D. Decreased lipolysis

24-27. All the following enzymes are activated by glucagon, *except:*
A. Phosphorylase kinase
B. Phosphoenol pyruvate carboxykinase
C. Liver phosphorylase
D. Phosphofructokinase

24-28. All are true with regard to the juvenile diabetes mellitus, *except:*
A. Patients need insulin injections
B. Frequent ketoacidosis is common
C. Insulin resistance is the cause
D. Genetic susceptibility is inherited

24-29. Which of the following laberatory results indicate poor glycemic control?

A. PPBS of 180 mg/dL
B. Serum cholesterol of 240 mg/dL
C. Glycohemoglobin of 10%
D. Blood urea level of 35 mg/dL

24-30. Which of the following enzymes is active in diabetes but not in starvation?

A. Hormone sensitive lipase
B. Pyruvate carboxylase
C. Carnitine acyl transferase
D. Aldose reductase

Answers (Chapter 24)

24-1. B.	24-2. D.	24-3. B.	24-4. A.
24-5. D.	24-6. B.	24-7. D.	24-8. A.
24-9. D.	24-10. A.	24-11. C.	24-12. C.
24-13. C	24-14. B	24-15. B.	24-16. D.
24-17. D.	24-18. C.	24-19. C.	24-20. C.
24-21. B.	24-22. D.	24-23. B.	24.24. D.
24-25. C.	24-26. A.	24-27. D.	24-28. C.
24-29. C	24-30. D.		

CHAPTER 25: CARDIOVASCULAR DISEASES

25-1. HDL cholesterol is said to be "good" cholesterol, because

A. HDL contains enzymes to breakdown cholesterol
B. HDL carries cholesterol from liver to tissues where it is broken down
C. HDL carries cholesterol from tissues to liver from where cholesterol is excreted
D. HDL inhibits cholesterol synthesis

25-2. All the following statements are true with regard to type IIA hyperlipoproteinemia, *except:*

A. Premature artherosclerosis
B. Elevated plasma LDL cholesterol
C. Creamy layer on top of the serum
D. Xanthomata

25-3. All are true in familial hyperlipoprotenemia Type II-A, *except:*
A. TAG level in blood is increased
B. LDL level in blood is increased
C. LDL receptor defect is seen
D. Cholesterol level in blood is increased

25-4. Normal blood cholesterol level is:
A. 40–6 mg/100 mL
B. 70–11 mg/100 mL
C. 120–15 mg/100 mL
D. 150–20 mg/100 mL

25-5. Hypercholesterolemia is seen in the following conditions, *except:*
A. Diabetes mellitus
B. Thyrotoxicosis
C. Nephrotic syndrome
D. Obstructive jaundice

25-6. The characteristic finding in hypobeta lipoproteinemia is:
A. Acanthocytosis
B. Orange tonsils
C. Eruptive xanthomas
D. Corneal arcus

25-7. Which of the following dietary modifications have minimum hypocholesterolemic action?
A. Inclusion of fresh fruits and vegetables in diet
B. Dietary cholesterol intake less than 300 mg/day
C. Intake of whole wheat bran
D. Consuming fish as the only non-vegetarian food

25-8. Premature atherogenesis will not set in when---- is defective.
A. Cholesterol efflux regulatory protein
B. LDL receptor
C. Cholesterol ester transfer protein
D. Apolipoprotein B

25-9. Hypoalpha lipoproteinemia is characteristic of:
A. Fish eye disease
B. Bassen Kornzweig syndrome
C. Floating beta disease
D. Tangier's disease

25-10. A family gives a history of several members completing a century of healthy life. The cause maybe:
A. Normal insulin secretion
B. Normal adipose tissue liver axis
C. Elevated HDL levels
D. Consumption of a fish based diet

25-11. Hypercholesterolemia in a 15-year-old girl may be due to the following causes, *except:*
A. Nephrotic syndrome
B. Hypothyroidism
C. Type IV hyperlipoproteinenmia
D. Obstructive jaundice

25-12. An obese person may have all the following biochemical abnormalities, *except*:
A. Increased glucose tolerance
B. Hypertriglyceridemia
C. Chronic respiratory acidosis
D. High plasma insulin levels

25-13. Which of the following is the most common type of hyperlipidemia in diabetes mellitus?
A. Elevated triglycerides with normal cholesterol
B. Normal triglycerides with elevated cholesterol
C. Elevated cholesterol and decreased triglycerides
D. Elevated cholesterol and triglycerides

25-14. The lipoprotein deficient in Tangier's disease
A. HDL B. LDL
C. Lp(a) D. IDL

25-15. Secondary hyperlipidemia occurs in all the following conditions, *except:*
A. Hypothyroidism
B. Alcoholism
C. Chronic pancreatitis
D. Nephrotic syndrome

Answers (Chapter 25)

25-1. C.	25-2. C.	25-3. A.	25-4. D.
25-5. B.	25-6. A.	25-7. C.	25-8. C.
25-9. D	25-10. C.	25-11. C.	25-12. A.
25-13. B	25-14. A.	25-15. C.	

CHAPTER 26: LIVER AND GASTRIC FUNCTION TESTS

26-1. All enzymes are elevated in obstructive liver disease, *except:*
A. Gamma glutamyl transferase (GGT)
B. 5' nucleotidase (NTP)
C. Alkaline phosphatase (ALP)
D. Lactate dehydrogenase (LDH)

26-2. All are features of obstructive jaundice, *except:*
A. Increased level of conjugated bilirubin in blood
B. Clay colored stools
C. Presence of bile salts in urine
D. Increased excretion of urobilinogen in urine

26-3. A patient with infective hepatitis is likely to have all the following findings, *except:*
A. Hyperbilirubinemia
B. Bilirubinuria
C. Absence of bile salts in urine
D. Elevated AST

26-4. An increase in serum unconjugated bilirubin occurs in:
A. Hemolytic jaundice
B. Obstructive jaundice
C. Defect in intestinal absorption
D. Glomerulonephritis

26-5. Conjugated hyperbilirubinemia with raised alkaline phosphatase levels are characteristic of:
A. Obstructive jaundice
B. Hemolytic jaundice
C. Viral hepatitis
D. Physiological jaundice

26-6. Gastric acid secretion is stimulated by all the following, *except:*
A. Hyperglycemia B. Gastrin
C. Vagus D. Histamine

26-7. Hypoacidity is found in all the following conditions *except:*
A. Pernicious anemia B. Carcinoma of stomach
C. Insulinoma D. Atrophic gastritis

26-8. All the following biochemical parameters are indices of liver function, *except:*

A. Bilirubin B. Cholesterol
C. Albumin D. Creatinine

26-9. All the following biochemical parameters are indices of liver function, *except:*

A. Bilirubin B. Cholesterol
C. Albumin D. Creatinine

26-10. A patient with infective hepatitis is likely to have all the following findings, *except:*

A. Hyperbilirubinemia
B. Bilirubinuria
C. Absence of bile salts in urine
D. Elevated AST

26.11. Which enzyme test is more specific for parenchymal (hepatocellular) liver disease?

A. Acid phosphatase (ACP)
B. Alanine aminotransferase (ALT)
C. Lactate dehydrogenase (LDH)
D. Amylase

26-12. Which hormone has no effect on gastric acid secretion?

A. Cholecystokinin B. Secretin
C. Gastrin D. Somatomedin

26-13. Which of the following tests is not indicated in a patient with generalized edema?

A. Liver function tests
B. Renal function tests
C. Thyroid function tests
D. Pancreatic function tests

26-14. Which of the following biochemical findings does not agree with acute hepatic failure?

A. Respiratory alkalosis
B. Hyperammonemia
C. Lactic acidosis
D. Uremia

26-15. The laboratory data that is against a diagnosis of obstructive jaundice is:

A. High alkaline phosphatse level
B. Increased excretion of urobilinogen in urine
C. Elevated serum cholesterol level
D. Direct positive van den Bergh reaction

Answers (Chapter 26)

26-1. D.	26-2. D.	26-3. B.	26-4. A.
26-5. A.	26-6. A.	26-7. C.	26-8. D.
26.9. D.	26.10. B.	26.11. B.	26-12. D.
26-13. D	26-14. D.	26-15. B.	

CHAPTER 27: KIDNEY FUNCTION TESTS

27-1. Normal specific gravity of urine is:
A. 1.003 to 1.010 B. 1.010 to 1.015
C. 1.015 to 1.025 D. 1.025 to 1.035

27-2. Polyuria is a characteristic feature of all the following, *except*:
A. Diabetes mellitus B. Glomerulo nephritis
C. Diabetes insipidus D. Chronic renal failure

27-3. Specific gravity of urine increases in:
A. Chronic glomerulonephritis
B. Diabetes mellitus
C. Liver diseases
D. Intake of vegetables

27-4. Proteinuria is seen in all the following conditions, *except:*
A. Acute glomerulonephritis
B. Diabetes insipidus
C. Pyelonephritis
D. Multiple myeloma

27-5. Which of the following compounds have a clearance value higher than GFR?
A. Uric acid B. Diodrast
C. Urea D. Inulin

27-6. Renal plasma flow is measured by:
A. Creatinine clearance
B. Inulin clearance
C. Para-aminohippurate clearance
D. Urine output

27-7. A patient with chronic renal failure will excrete:
A. Acid urine
B. Hypertonic urine
C. Urine with specific gravity more than 1.18
D. Urine with a high creatinine concentration

27-8. Which indicates an abnormal renal function?
A. Blood urea 30 mg/dL
B. GFR = 125 ml/min
C. Serum creatinine 8 mg/dL
D. Urine pH = 6.8

27-9. Urea clearance is lowered in:
A. Acute nephritis
B. Chronic liver failure
C. Chronic cardiac failure
D. Benign hypertension

27-10. Normal value of blood urea is:
A. 3–4 mg/100 mL B. 4–8 mg/100 mL
C. 8–16 mg/100 mL D. 20–40 mg/100 mL

27-11. Failure of concentrating capacity of urine is assessed by:
A. Urea clearance
B. Water deprivation test
C. Specific gravity of urine after giving water
D. Creatinine level in urine

27-12. Urine of a patient with obstructive jaundice will give a positive test for all the following, *except:*
A. Gmelin's test B. Hay's test
C. Fouchet's test D. Benzidine test

27-13. Urine will have abnormal color in all the following conditions, *except:*
A. Melanuria B. Porphyria
C. Creatinuria D. Alkaptonuria

27-14. All the following statements regarding glomerular filtration is correct, *except:*
A. GFR is approximately equal to endogenous creatinine clearance
B. Measurement of GFR is a test of renal tubular function
C. Glomerular filtrate has a specific gravity of 1.010
D. Glomerular filtrate is formed by ultrafiltration of plasma

27-15. Clearance values of all the following compounds are used to test renal function, *except:*
A. Creatinine B. Inulin
C. Glucose D. Urea

27-16. When 15N (radioactive nitrogen) labeled glycine is given, the label will appear in all the following compounds in urine, *except:*

A. Urobilinogen B. Urea
C. Urocanic acid D. Uric acid

27-17. Normal urine contains all following, *except:*

A. Glucose B. Ethereal sulfate
C. Bicarbonate D. Creatinine

27-18. Opalescent urine is seen in:

A. Porphyria B. Alkaptonuria
C. Chyluria D. Creatinuria

27-19. Which anions are present in normal urine?

A. Phosphate B. Sulfate
C. Chloride D. Proteinate

27-20. Specific gravity of urine increases in:

A. Chronic glomerulonephritis
B. Diabetes mellitus
C. Liver diseases
D. Intake of vegetables

27-21. A patient with chronic renal failure may have all the following features, *except:*

A. Urine with fixed specific gravity of 1.010
B. Protein creatinine ratio more than 3
C. Positive heat and acetic acid test
D. Nocturnal polyuria

27-22. All the following tests are answered by normal urine, *except:*

A. Ehrlich's test B. Rothera's test
C. Jaffe's test D. Schlesinger's test

27-23. Which of the following constituents of urine does not contain nitrogen?

A. Urobilinogen B. Uric acid
C. Acetoacetic acid D. Urea

27-24. To calculate the creatinine clearance, all the following data are required, *except:*

A. Plasma creatinine level
B. Patient's height and weight
C. Daily urine output
D. Specific gravity of urine

27-25. Urea level in blood may be elevated in all the following conditions, *except:*

A. Dehydration B. Renal failure
C. Hepatic failure D. Hypertension

27-26. Renal stones may contain all the following organic constituents, *except*:

A. Cystine B. Uric acid
C. Cholesterol D. Xanthine

27-27. All the following laboratory data are suggestive of acute renal failure, *except:*

A. Plasma sodium—150 mmol/L
B. Plasma potassium—5.6 mmol/L
C. Urea—220 mg/dL
D. Creatinine—3.2 mg/dL

27-28. Which ion is decreased in renal failure:

A. Phosphate B. Potassium
C. Bicarbonate D. Hydrogen ions

27-29. Renal tubular function can be assessed by:

A. Urea clearance
B. Plasma electrolytes
C. Specific gravity of early morning sample of urine
D. Plasma urea

27-30. Plasma urea is increased but creatinine is normal in all the following, *except:*

A. High protein intake
B. Gastrointestinal bleeding
C. Starvation
D. Muscle wasting

Answers (Chapter 27)

27-1. C.	27-2. B.	27-3. B.	27-4. B.
27-5. B.	27-6. C.	27-7. D.	27-8. C.
27-9. A.	27-10. D.	27-11. B.	27-12. D.
27-13. C.	27-14. B.	27-15. C.	27-16. C.
27-17. A.	27-18. C.	27-19. D.	27-20. B.
27-21. C.	27-22. B.	27-23. C.	27-24. D.
27-25. C.	27-26. C.	27-27. A.	27-28. C
27-29. C.	27-30. D.		

CHAPTER 28: PLASMA PROTEINS

28-1. Normal level of albumin in blood is:

A. 1.5–2.5 mg/dL B. 2.5–3.5 mg/dL
C. 2.5–3.5 g/dL D. 3.5–5.0 g/dL

28-2. Hypoalbuminemia is seen in all the following conditions, *except*:

A. Cirrhosis of liver B. Nephrotic syndrome
C. Malnutrition D. Acute infections

28-3. Hemopexin carries:

A. Free hemoglobin B. Free heme
C. Free bilirubin D. Free iron

28-4. All the following are acute phase reactant proteins, *except*:

A. C-reactive protein (CRP)
B. HDL
C. Ceruloplasmin
D. Haptoglobin

28-5. Polymorphism is exhibited by all the following proteins, *except:*

A. Haptoglobin B. Transferrin
C. Albumin D. Ceruloplasmin

28-6. Wilson's hepatolenticular degeneration is characterized by:

A. Ceruloplasmin level in blood is increased
B. Copper is accumulated in liver to produce cirrhosis
C. Copper is deposited in skin to produce bronze color
D. Autosomal dominant inheritance

28-7. All are true with regard to alpha-1 antitrypsin (AAT), *except:*

A. It is a protease inhibitor
B. It shows polymorphism
C. Deficiency leads to emphysema in lungs
D. Deficiency is associated with edema

28-8. Albumin level in blood is estimated by:

A. Jaffe's picric acid reaction
B. Bromocresol green reaction
C. Diacetyl monoxime method
D. Chromatography

28-9. Which reaction takes place exclusively in liver?

A. Gluconeogenesis
B. Glycolysis
C. Glycogen synthesis
D. Albumin synthesis

28-10. In blood, all are bound with albumin, *except:*

A. Non-esterified fatty acids
B. Bilirubin
C. Iron
D. Salicylate

28-11. Electrophoretic separation of proteins is of diagnostic value in all conditions, *except:*

A. Nephrotic syndrome
B. Multiple myeloma
C. Alpha-1 antitrypsin deficiency
D. Acquired immunodeficiency syndrome

28-12. Edema due to hypoproteinemia may be seen in all the following clinical conditions, *except:*

A. Rheumatoid arthritis
B. Cirrhosis liver
C. Malnutrition
D. Nephrotic syndrome

28-13. Albumin/globulin ratio is reversed in all the following conditions, *except:*

A. Cirrhosis liver
B. Primary immune deficiency
C. Nephrotic syndrome
D. Multiple myeloma

28-14. It is not safe to give salicylate to infants having hemolytic disease, because:

A. Bilirubin and salicylate compete each other for binding on albumin
B. Salicylate is toxic to children
C. Hemolysis is produced by salicylate
D. Both bilirubin and salicylate are metabolized by the same enzyme systems in liver

28-15. All are transport proteins of blood, *except:*

A. Albumin B. Transcortin
C. Transferrin D. Ceruloplasmin

28-16. One of the plasma proteins listed below is not a transport protein (carrier protein)

A. Transferrin
B. Haptoglobin
C. Albumin
D. Alpha-1-antitrypsin

28-17. All are true with ceruloplasmin, *except:*

A. It is a copper containing protein found in serum
B. It is an acute phase protein
C. Blood level is increased in Wilson's disease
D. It has ferroxidase activity

28-18. All are correct with hemophilia, *except:*

A. Inherited as an X-linked recessive trait
B. Inherited by son from father
C. Daughters are likely to be carriers
D. Prenatal diagnosis is important

28-19. Hemorrhagic tendency is observed in all the following conditions, *except*:

A. Hemophilia
B. Vitamin K deficiency
C. Thrombocytopenia
D. Phenylketonuria

28-20. Which does not show X-linked inheritance:

A. Glucose-6-P dehydrogenase deficiency
B. Hemophilia
C. Von Willebrandt's disease
D. Christmas disease

Answers (Chapter 28)

28-1. D.	28-2. D.	28-3. B.	28-4. B.
28-5. C.	28-6. B.	28-7. D.	28.8. B.
28-9. D.	28-10. C.	28-11. D.	28-12. A
28-13. B.	28-14. A.	28-15. D.	28-16. D.
28-17. C.	28-18. B.	28-19. D.	28-20. C.

CHAPTER 29: ACID-BASE BALANCE AND pH

29-1. Causes of metabolic alkalosis include the following, *except:*

A. Prolonged vomiting
B. Continuous gastric aspiration
C. Hyperventilation
D. Ingestion of antacids

29-2. Biochemical findings in diabetic ketoacidosis include the following, *except:*

A. pCO_2 is lowered or normal
B. Increased anion gap
C. Lowered bicarbonate level
D. Elevation of plasma chloride

29-3. In respiratory acidosis, patient will show which of the following features?

A. a low pCO_2
B. Lowered plasma bicarbonate
C. Reduced rate and depth of respiration
D. Decreased hydrogen ion concentration in plasma

29-4. A patient treated with carbonic anhydrase inhibitor may develop:

A. Metabolic acidosis
B. Respiratory acidosis
C. Respiratory alkalosis
D. Metabolic alkalosis

29-5. Administration of diuretics cause loss of potassium, which may lead to:

A. Metabolic acidosis
B. Respiratory acidosis
C. Respiratory alkalosis
D. Metabolic alkalosis

29-6. Patients with diarrhea may develop the following complications, *except:*

A. Metabolic alkalosis B. Hypertonic contraction
C. Dehydration D. Metabolic acidosis

29-7. Which of the following conditions will produce a high anion gap acidosis?

A. Diarrhea B. Renal tubular acidosis
C. Renal failure D. Ureterosigmoidostomy

29-8. Metabolic alkalosis is characterized by:

A. High hydrogen ion concentration in plasma
B. Increased pCO_2
C. Elevated chloride level in serum
D. Low potassium level

29-9. Respiratory acidosis can result from all the following, *except:*

A. Bronchial asthma B. Narcotic poisoning
C. Poliomyelitis D. Artificial ventilation

29-10. Non-respiratory alkalosis can result from:
A. Persistent vomiting
B. Salicylate poisoning
C. Chronic renal failure
D. Cyanotic heart disease

29-11. The biochemical findings in diabetic ketoacidosis include the following, *except:*
A. pCO_2 is lowered
B. Increased anion gap
C. Lowered bicarbonate
D. Elevation of chloride

29-12. ABG data: pH—7.47, pCO_2—30 mm of Hg, Bicarbonate—19 mmol/L. This suggests what?
A. Acute respiratory alkalosis
B. Partially compensated respiratory alkalosis
C. Partially compensated metabolic acidosis
D. Mixed respiratory and metabolic alkalosis

29-13. In the blood (pH = 7.4), which of the following buffer system is most effective?
A. Citric acid buffer (pK = 3.0)
D. Glutamate buffer (pK = 3.2)
B. Carbonic acid buffer (pK = 6.1)
C. Phosphate buffer (pK = 6.8)

29-14. Identify acid-base disturbance in this patient: pH 7.56, pCO_2—54 mm of Hg, Bicarbonate—45 mmol/L.
A. Chronic respiratory acidosis
B. Mixed respiratory acidosis and alkalosis
C. Partially compensated metabolic alkalosis
D. Uncompensated respiratory acidosis

29-15. In respiratory acidosis, patient will show which of the following features?
A. Low pCO_2
B. Lowered plasma bicarbonate
C. Reduced rate and depth of respiration
D. Decreased hydrogen ion concentration in plasma

29-16. All are causes of metabolic acidosis, *except:*
A. Ureterosigmoidostomy
B. Ingestion of antacids
C. Addison's disease
D. Diabetes mellitus

29-17. When pH falls by 1 unit, the hydrogen ion concentration:
A. Decreases 10 times
B. Increases two fold
C. Changes by 7 times
D. Increases 10 times

29-18. A patient with COPD (chronic obstructive pulmonary disease) on loop diuretics may have:
A. Respiratory and metabolic acidosis
B. Respiratory and metabolic alkalosis
C. Respiratory acidosis and metabolic alkalosis
D. Respiratory alkalosis and metabolic acidosis

29-19. In a case of partially compensated respiratory acidosis, the findings in plasma include:
A. Normal pCO_2 level
B. Elevated bicarbonate level
C. Increased pH
D. Bicarbonate to carbonic acid ratio is 20

29-20. Overcorrection of acidosis can cause all the following complications, *except:*
A. Shift of oxygen dissociation curve (ODC) to left
B. Sudden hypoxia
C. Hyperkalemia
D. Decrease in hydrogen ions

29-21. In compensated metabolic alkalosis,
A. Rrespiratory center is stimulated
B. pCO_2 increases
C. pO_2 elevated
D. Urine becomes acidic

29-22. Rapid correction of acidosis in a patient with hypoxia leads to:
A. Improvement of pO_2
B. Decreased tissue oxygenation
C. Decreased affinity of hemoglobin for oxygen
D. Exaggerated 2,3 BPG effect

29-23. Which of the following parameters indicate compensated state in a case of respiratory alkalosis:
A. High pCO_2
B. Low anion gap
C. Low bicarbonate
D. Low potassium level

29-24. Proximal renal tubular acidosis is characterized by all, *except:*
A. Metabolic acidosis B. Normal anion gap
C. Urine pH of 4.5 D. Hyperchloremia

Answers (Chapter 29)

29-1. C.	29-2. D.	29-3. C.	29-4. A.
29-5. D.	29-6. A.	29-7. C.	29-8. D.
29-9. D.	29-10. A.	29-11. D.	29-12. B.
29-13. B.	29-14. C.	29-15. C.	29-16. B.
29-17. D.	29-18. C.	29-19. B.	29-20. C.
29-21. B.	29-22. B.	29-23. B.	29-24. C.

CHAPTER 30: ELECTROLYTES

30-1. The intercellular cation present in maximum concentration is:
A. Potassium B. Magnesium
C. Sodium D. Calcium

30-2. The extracellular cation present in maximum concentration is:
A. Potassium B. Magnesium
C. Sodium D. Calcium

30-3. When there is deficiency of ADH (antidiuretic hormone):
A. ECF volume expands
B. Plasma osmolarity increases
C. Sodium depletion occurs
D. Thirst sensation is suppressed

30-4. Hypotonic expansion of extracellular fluid occurs:
A. Hyperaldosteronism
B. Inappropriate secretion of ADH
C. Cushing's syndrome
D. Intravenous infusion with normal saline

30-5. All the following conditions produce isotonic expansion of extracellular fluid volume, *except:*
A. Congestive cardiac failure
B. Infusion with normal saline
C. Hyperaldosteronism
D. Pulmonary edema

30-6. Which of the following parameters are important in the assessment of fluid balance, *except:*
A. Hematocrit value
B. Osmolality of urine
C. Specific gravity of urine
D. Urinary pH

30-7. All the following hormones affect fluid and electrolyte balance, *except:*
A. Aldosterone B. Anti diuretic hormone
C. Cortisone D. Thyroxine

30-8. The type of fluid electrolyte imbalance seen in a case of diabetes insipidus is:
A. Isotonic expansion
B. Hypotonic contraction
C. Hypertonic contraction
D. Hypertonic expansion

30-9. From pairs of diseases and associated abnormalities, pick out the mismatched pair:
A. Maple syrup urine disease and metabolic acidosis
B. Conn's syndrome and metabolic alkalosis
C. SIADH and hypertonic expansion
D. Waldenstroem's macroglobulinemia and hyperviscosity

30-10. The major difference between intracellular fluid (ICF) and extracellular fluid (ECF) is that:
A. ECF has a lower protein concentration than ICF
B. Chloride level is higher in ECF
C. Sodium concentration is lower in ECF
D. Phosphates form the major anion of ECF

30-11. Secretion of ADH is stimulated by all, *except:*
A. Increased sodium in serum
B. High osmolality of plasma
C. Inability to retain sodium
D. Contraction of ECF volume

30-12. A patient with diarrhea may have all the following abnormalities, *except:*
A. Metabolic acidosis
B. Isotonic contraction of ECF
C. Hypertonic contraction of ECF
D. Urine with a high specific gravity

Answers (Chapter 30)

30-1. A. 30-2. C. 30-3. C. 30-4. B.
30-5. D. 30-6. A. 30-7. D. 30-8. C.
30-9. C. 30-10. A. 30-11. C. 30-12. B.

CHAPTER 34: CLINICAL LABORATORY

34-1. Which is the anticoagulant of choice for biochemical estimations in general:
A. EDTA B. Oxalate–flouride mixture
C. Lithium heparin D. Citrate

34-2. Which of the following is not a point of care test (POCT):
A. ABG analysis B. Plasma electrolytes
C. Plasma glucose D. Plasma proteins

34-3. Which of the following is not an emergency investigation?
A. Potassium B. Fasting glucose
C. Creatinine D. Calcium

34-4. Which of the following parameters is not affected even if the assay is carried out after 6 hours of collection:
A. Glucose B. Potassium
C. Carbondioxide D. Albumin

34-5. Which of the following parameters is not affected by using a tourniquet while collecting blood?
A. Calcium B. Ammonia
C. Glucose D. Lactate

Answers (Chapter 34)

34-1. C. 34-2. D. 34-3. B. 34-4. D.
34-5. C.

CHAPTER 35: LABORATORY TECHNIQUES

35-1. Mobility of particles during electrophoresis is based on all the following factors, *except:*
A. Current voltage and ampere
B. Quality of supporting medium
C. pH of buffer medium
D. Solubility of particular protein

35-2. A polysaccharide made up of D-galactose and anhydrogalactose units, which are used as matrix for electrophoresis:

A. Cellobiose
B. Dextran
C. Polystyrene
D. Agarose

35-3. In a protein molecule, normal glutamic acid is substituted to valine. Both normal and mutated proteins are electrophoresed at pH 8.6. The mobility will be:

A. Mutated protein moves faster than normal protein towards positive pole
B. Mutated protein moves slower than normal protein towards positive pole
C. Mutated protein remains at the point of application
D. Mutated protein moves to negative pole and normal protein to positive pole

35-4. The quickest method for separation of protein:

A. Electrophoresis
B. High performance liquid chromatography
C. Ion exchange chromatography
D. Thin layer chromatography

35-5. Molecular weight of a protein may be calculated by all the following methods, *except:*

A. Ultra centrifugation
B. Gel filtration
C. Polyacrylamide gel electrophoresis
D. Affinity chromatography

35-6. Ultracentrifugation is useful for all the following, *except:*

A. Isolation of mitochondria
B. Preparation of blood group antigens
C. Identification of lipoproteins
D. Separation of proteins

35-7. The chromogens used in ELISA test include all the following, *except:*

A. DAB (diamino benzidine)
B. DMAB (dimethyl amino azo benzene)
C. NBT (nitro blue tetrazolium)
D. NPP (nitro phenyl phosphate)

35-8. Which of the following has maximum absorption at 340 nm?

A. Heme B. Proteins
C. Nucleic acid D. NADH

35-9. The electrolytes such as Na^+ and K^+ can be measured by:

A. Flame photometer
B. Spectrophotometer
C. Radioimmunoassay
D. High performance liquid chromatography

35-10. All the techniques listed below measures proteins, *except:*

A. Nephelometry B. Flame photmetry
C. ELISA D. Radial immunodiffusion

35-11. Immunosorbent assays will use antibody conjugated to all the following reagents, *except*:

A. Alkaline phosphatase (ALP)
B. Horse radish peroxidase (HRP)
C. Riboflavin
D. Biotin

Answers (Chapter 35)

35-1. D.	35-2. D.	35-3. B.	35-4. B.
35-5. D.	35-6. B.	35-7. B.	35-8. D.
35-9. A.	35-10. B.	35-11. C.	

CHAPTER 36: FAT SOLUBLE VITAMINS (A,D,E,K)

36-1. Daily requirement of vitamin A for a normal adult is:

A. 1 µg B. 5 µg
C. 100 µg D. 750 µg

36-2. Deficiency of vitamin A leads to:

A. Night blindness
B. Rickets
C. Macrocytic anemia
D. Microcytic anemia

36-3. Nyctalopia is due to the deficiency of:

A. Vitamin K B. Vitamin E
C. Vitamin B_{12} D. Vitamin A

36-4. When kidney diseases are present, oral doses of vitamin D may not be effective in curing rickets, because:

A. Hydroxylation reaction is taking place in kidney, which activates vitamin
B. Dehydrogenation of vitamin D is taking place in kidney
C. Hydroxylation of vitamin D is taking place in kidney which destroys vitamin
D. Vitamin D is stored in liver

36-5. Daily requirement of vitamin D is:

A. 1 μg
B. 10 μg
C. 100 μg
D. 750 μg

36-6. Deficiency of vitamin D leads to:

A. Night blindness
B. Macrocytic anemia
C. Rickets
D. Microcytic anemia

36-7. Cholecalciferol is synthesized in:

A. Liver
B. Skin
C. Kidney
D. Intestinal mucosa

36-8. The most important biological role for vitamin E is:

A. To produce clotting factors
B. Antidote of selenium poisoning
C. Anticoagulant
D. Antioxidant

36-9. Large doses of vitamin K in neonates may cause:

A. Rebound bleeding
B. Porphyria
C. Jaundice
D. Cyanosis

36-10. Deficiency of vitamin K can occur in the following conditions, *except:*

A. Following gastrectomy
B. Prolonged antibiotic therapy
C. Obstructive jaundice
D. Administration of dicoumarol

36-11. Biochemical function of vitamin K is for:

A. Converting proline to hydroxyproline
B. Conversion of prothrombin to thrombin
C. Gamma carboxylation of clotting factors
D. Inhibition of lipid peroxidation in biomembranes

36-12. Vitamin K is inhibited by:
A. Isoniazid (INH) B. Methotrexate
C. Dicoumarol D. Avidin

36-13. Skin and mucous membrane are affected in deficiency of all the vitamins listed below, *except:*
A. Retinol B. Niacin
C. Riboflavin D. Calcitriol

36-14. Deficiency of vitamin K can occur in the following conditions, *except:*
A. Following gastrectomy
B. Prolonged antibiotic therapy
C. Obstructive jaundice
D. Administration of dicoumarol

36-15. The structure of vitamin A contains:
A. Chromane ring
B. Beta ionone ring
C. Thiazole ring
D. Naphthoquinone ring

36-16. The structure of vitamin E contains:
A. Chromane ring
B. Beta ionone ring
C. Thiazole ring
D. Naphthoquinone ring

36-17. All are good sources of vitamin A, *except:*
A. Pumpkin B. Carrot
C. Mangoes D. Oranges

36-18. Rickets may occur in all conditions, *except:*
A. Chronic renal failure
B. Liver diseases
C. Under exposure to sun light
D. Prolonged antibiotic therapy

36-19. Cholecalciferol is synthesized in:
A. Liver B. Skin
C. Kidney D. Intestinal mucosa

36-20. Which is the enzyme specially involved in ossification of bone?
A. Alkaline phosphatase
B. Acid phosphatase
C. Hexokinase
D. ATPase

36-21. When kidney diseases are present, oral doses of vitamin D may not be effective in curing rickets, because vitamin D is:

A. Stored in kidneys
B. Dehydrogenated in kidneys
C. Destroyed in kidneys
D. Activated in kidney

36-22. If vitamin K is given to a patient with hemophilia:

A. Bleeding time prolonged, clotting time remains normal
B. Blotting time prolonged, bleeding time remains normal
C. Both bleeding and clotting times prolonged
D. Both bleeding and clotting times decreased

36-23. Excess vitamin K in neonates may cause:

A. Rebound bleeding B. Porphyria
C. Jaundice D. Cyanosis

36-24. All these vitamins are required for normal bone formation, *except:*

A. Vitamin C B. Vitamin D
C. Vitamin A D. Vitamin E

36-25. Bleeding tendency is common in deficiency of all the following, *except:*

A. Vitamin K B. Vitamin B_{12}
C. Vitamin C D. Platelets

36-26. The active form of vitamin D is known as:

A. Cholecalciferol
B. Cyclopentano phenanthrene
C. Calcitriol
D. Lanosterol

36-27. Richest source of vitamin D is:

A. Fresh leafy vegetables
B. Fish liver oil
C. Egg yolk
D. Vegetable oils

36-28. All water soluble vitamins are excreted through urine and are not stored in the body, *except:*

A. vitamin B_{12} B. vitamin B_6 (pyridoxal)
C. vitamin C D. Niacin

Answers (Chapter 36)

36-1. D.	36-2. A.	36-3. D.	36-4. A.
36-5. B.	36-6. C.	36-7. B.	36-8. D.
36-9. C.	36-10. A.	36-11. C.	36-12. C.
36-13. D	36-14. A.	36-15. B.	36-16. A.
36-17. D.	36-18. D.	36-19. B.	36-20. A.
36-21. D.	36-22. D.	36-23. C.	36-24. D.
36-25. B.	36-26. C.	36-27. B.	36-28. A.

CHAPTER 37: WATER SOLUBLE VITAMINS

37-1. In thiamine deficiency, serum levels of the following compound is increased:

A. Glycine B. Pyruvic acid
C. Glutamic acid D. Transketolase

37-2. Thiamine deficiency is first manifested as:

A. Reduced transketolase activity in RBC
B. Increased glucose-6-phosphate dehydrogenase activity
C. Absence of aldolase in RBC
D. Altered lactate dehydrogenase isoenzymes in RBC

37-3. Beriberi is due to the deficiency of:

A. Niacin B. Thiamine
C. Riboflavin D. Vitamin B_{12}

37-4. Daily requirement of vitamin B_1 (thiamine) for a normal healthy adult is

A. 1 μg B. 5 μg
C. 100 μg D. 1 μg

37-5. All the following vitamins are produced from precursors in the human body, *except:*

A. Niacin B. Ascorbic acid
C. Vitamin D D. Vitamin A

37-6. Neuritis is a manifestation of the deficiency of all the following vitamins, *except:*

A. Thiamine B. Vitamin B_{12}
C. Pantothenic acid D. Vitamin C

37-7. Which vitamin is required for oxidative decarboxylation?

A. Pyridoxal phosphate B. Thiamine
C. Biotin D. Riboflavin

37-8. Glossitis is due to the deficiency of:

A. Niacin B. Thiamine

C. Riboflavin D. Vitamin B_{12}

37-9. All the following are riboflavin (FAD) dependent reactions, except:

A. Succinate dehydrogenase

B. Acyl CoA dehydrogenase

C. Xanthine oxidase

D. Lactate dehydrogenase

37-10. The water soluble vitamin synthesized in human body is:

A. Ascorbic acid B. Pyridoxine

C. Niacin D. Pantothenic acid

37-11. Pellagra is due to the deficiency of

A. Niacin B. Thiamine

C. Riboflavin D. Vitamin B_{12}

37-12. All the following conditions can lead to symptoms of pellagra, *except:*

A. Low tryptophan content in diet

B. Anti/tuberculosis therapy by isoniazid

C. Excessive intake of alcohol

D. High leucine content of diet

37-13. Deficiency of vitamin B_6 is manifested as:

A. Wet beriberi B. Scurvy

C. Rickets D. Infantile convulsions

37-14. Daily requirement of pyridoxine for an adult is

A. 2 mg B. 10 mg

C. 20 mg D. 70 mg

37-15. Pyridoxal phosphate is required for all the following reactions, *except:*

A. Transamination

B. Deamination

C. Decarboxylation of amino acids

D. Decarboxylation

37-16. Deficiency of pantothenic acid leads to:

A. Night blindness B. Rickets

C. Macrocytic anemia D. Burning foot syndrome

37-17. Biotin is inhibited by:

A. Isoniazid (INH) B. Methotrexate

C. Dicoumarol D. Avidin

37-18. Biotin is a co-enzyme for all the following carbon dioxide fixation reactions, *except:*

A. Pyruvate carboxylase
B. Acetyl CoA carboxylase
C. Propionate carboxylase
D. Carbamoyl phosphate synthetase

37-19. All compounds act as antivitamin, *except:*

A. Avidin
B. Menadione
C. INH
D. Methotrexate

37-20. Riboflavin is a constituent of:

A. Co-carboxylase
B. Co-decarboxylase
C. NAD+
D. FAD

37-21. Vitamin B_6 is required for:

A. Conversion of glycine to serine
B. Production of gamma amino butyric acid (GABA)
C. Production of xanthurenic acid
D. Esterification of palmitic acid

37-22. Pantothenic acid is the co-enzyme for:

A. Decarboxylation of glutamic acid
B. Formation of fumarate
C. Oxidation of pyruvic acid
D. Methylation of homocysteine

37-23. Xanthurenic acid is excreted in urine in the deficiency of

A. Folic acid
B. Vitamin B_{12}
C. Niacin
D. Pyridoxal phosphate

37-24. Chronic alcoholism or consumption of milled rice may result in deficiency of:

A. Vitamin B_6
B. Vitamin B_{12}
C. Vitamin K
D. Vitamin B_1

37-25. All are features of pellagra, *except:*

A. Skin lesions
B. Break down of scars
C. Diarrhea
D. Decreased mental activity

37-26. Pyridoxine deficiency will affect the formation of all the following, *except:*

A. Gamma amino butyric acid
B. Glutamine
C. Dopamine
D. Serotonin

37-27. Transamination reactions require:
A. Pyridoxal phosphate
B. Vitamin B_{12}
C. Thiamine
D. Vitamin C

37-28. All are NAD+ dependent enzymes, *except:*
A. Lactate dehydrogenase
B. Glyceraldehyde-3-phosphate dehydrogenase
C. Pyruvate dehydrogenase
D. Glucose-6-phosphate dehydrogenase

37-29. All are NADP+ dependent enzymes, *except:*
A. Glucose-6-phosphate dehydrogenase
B. Cytoplasmic isocitrate dehydrogenase
C. 6-phosphogluconate dehydrogenase
D. Glycerophosphate dehydrogenase

37-30. Which of the following vitamins is required for oxidative decarboxylation?
A. Pyridoxal phosphate
B. Thiamine pyrophosphate
C. Biotin
D. Riboflavin

Answers (Chapter 37)

37-1. B.	37-2. A.	37-3. B.	37-4. D.
37-5. B.	37-6. D.	37-7. B.	37-8. C.
37-9. D.	37-10. C.	37-11. A.	37-12. C.
37-13. D.	37-14. A.	37-15. D.	37-16. D.
37-17. D.	37-18. D.	37-19. B.	37-20. D.
37-21. B.	37-22. C.	37-23. D.	37-24. D.
37-25. B.	37-26. B.	37-27. A.	37-28. D.
37-29. D.	37-30. B.		

CHAPTER 38: WATER SOLUBLE VITAMINS CONTINUED

38-1. Folic acid is inhibited by:
A. Isoniazid (INH)
B. Methotrexate
C. Dicoumarol
D. Avidin

38-2. Daily requirement of folic acid is:
A. 1 μg
B. 5 μg
C. 100 μg
D. 750 μg

38-3. Deficiency of folic acid leads to:
A. Night blindness B. Rickets
C. Macrocytic anemia D. Microcytic anemia

38-4. In vitamin B_{12} deficiency, which compound is excreted in urine?
A. FIGLU (formimino glutamic acid).
B. Homocysteine
C. Xanthurenic acid
D. Methyl malonic acid

38-5. Formimino glutamic acid (FIGLU) is excreted in urine in the deficiency of:
A. Folic acid B. Vitamin B_{12}
C. Niacin D. Pyridoxal phosphate

38-6. Daily requirement of vitamin B_{12} for a normal healthy adult is:
A. 1 μg B. 5 μg
C. 100 μg D. 750 μg

38-7. The metal present in vitamin B_{12} is:
A. Copper B. Cobalt
C. Chromium D. Manganese

38-8. Daily requirement of ascorbic acid for an adult:
A. 2 mg B. 10 mg
C. 20 mg D. 70 mg

38-9. Scurvy is due to the deficiency of:
A. Niacin B. Thiamine
C. Riboflavin D. Vitamin C

38-10. Pernicious anemia is caused by:
A. Inadequate intake of vitamin B_1
B. Absence of hydrochloric acid in gastric juice
C. Absence of intrinsic factor of the gastric juice
D. Over production of extrinsic factor

38-11. One of the features listed below is not true regarding pernicious anemia:
A. Neurological manifestations
B. Megaloblastic anemia
C. Increased gastrin secretion
D. Decreased basal acid output

38-12. Methotrexate is acting by inhibition of"
A. Cellular oxidation B. RNA synthesis
C. Cyanocobalamine D. Folate reductase

38-13. The structure of vitamin B_{12} contains:
A. Pyridine ring B. Corrin ring
C. Imidazole ring D. Pteridine ring

38-14. Megaloblastic anemia is due to deficiency of:
A. Niacin B. Thiamine
C. Riboflavin D. Vitamin B_{12}

38-15. Which acts as a strong reducing agent?
A. Folic acid B. Tocopherol
C. Ascorbic acid D. Pantothenic acid

38-16. A patient with vitamin C deficiency will show all the following clinical features, *except:*
A. Bone pains
B. Delayed wound healing
C. Subcutaneous hemorrhages
D. Megaloblastic anemia

38-17. A patient who has undergone gastrectomy is likely to develop deficiency of:
A. Vitamin A B. Vitamin B_{12}
C. Vitamin B_1 D. Vitamin K

38-18. Absorption of vitamin B_{12} takes place from
A. Stomach B. Duodenum
C. Jejunum D. Ileum

38-19. All the following vitamins are produced from precursors in the human body, *except:*
A. Niacin B. Ascorbic acid
C. Vitamin D D. Vitamin A

38-20. Pernicious anemia is characterized by all the following, *except:*
A. Histamine fast achlorhydria
B. Megaloblastic anemia
C. Subacute combined degeneration of spinal cord
D. Response to oral administration of folic acid

38-21. Neuritis is a manifestation of the deficiency of all the following vitamins, *except:*
A. Thiamine B. Pyridoxal
C. Vitamin B_{12} D. Vitamin C

38-22. The vitamin totally absent in plant source is:
A. Thiamine B. Vitamin B_{12}
C. Riboflavin D. Vitamin K

38-23. All substances contain sulfur, *except:*

A. Thiamine B. Tocopherol
C. Biotin D. Taurine

38-24. Folic acid is required for synthesis of:

A. ATP B. TTP
C. GTP D. CTP

38-25. All manifestations are seen in vitamin B_6 (pyridoxal phosphate) deficiency, *except:*

A. Skin rashes as in pellagra
B. Macrocytic anemia
C. Peripheral neuritis
D. Convulsions in children

38-26. Cystathionuria may be seen in the deficiency of all the following, *except:*

A. Vitamin B_{12}
B. Folic acid
C. Pyridoxal phosphate
D. Vitamin K

38-27. Riboflavin is needed as a co-enzyme (FAD) in the reaction:

A. Glyceraldehyde-3-phosphate dehydrogenase
B. Glucose-6-phosphate dehydrogenase
C. Fatty acyl CoA dehydrogenase
D. Malate dehydrogenase

38-28. In human body, highest concentration of ascorbic acid is found in:

A. Liver
B. Adrenal cortex
C. Posterior pituitary gland
D. Muscle

Answers (Chapter 38)

38-1. B.	38-2. C.	38-3. C.	38-4. D.
38-5. A.	38-6. A.	38-7. B.	38-8. D.
38-9. D.	38-10. C.	38-11. C.	38-12. D.
38-13. B.	38-14. D.	38-15. C.	38-16. D.
38-17. B.	38-18. D.	38-19. B.	38-20. D.
38-21. D.	38-22. B.	38-23. B.	38-24. D.
38-25. B.	38-26. D.	38-27. C.	38-28. B.

CHAPTER 39: MINERAL METABOLISM

39-1. Absorption of calcium from intestine is influenced by all the following, *except:*

A. 1,25-dihydrocholecalciferol
B. Calcitonin
C. Parathyroid hormone
D. Phosphate content of diet

39-2. Plasma calcium level is influenced by the following, *except:*

A. Calcitriol
B. Parathyroid hormone
C. Calcitonin
D. Thyroxin

39-3. Normal level of calcium in blood is:

A. 3–4 mg/dL
B. 9–11 mg/dL
C. 4–5 mEq/L
D. 96–106 mEq/L

39-4. Daily requirement of calcium for an adult is:

A. 2 mg
B. 15 mg
C. 70 mg
D. 500 mg

39-5. Calcium is required for the following, *except:*

A. Coagulation
B. Neuromuscular transmission
C. Absorption of iron
D. Intracellular messenger

39-6. Sodium level in serum is affected by all the following factors, *except:*

A. Aldosterone
B. Cortisone
C. ADH
D. Cholecalciferol

39-7. Normal level of sodium in plasma is:

A. 3–4 mg/dL
B. 9–11 mg/dL
C. 4–5 mEq/L
D. 136–145 mEq/L

39-8. Normal level of potassium in plasma is:

A. 3–4 mg/dL
B. 9–11 mg/dL
C. 4–5 mEq/L
D. 96–106 mEq/L

39-9. Normal level of chloride in plasma is:

A. 3–4 mg/dL
B. 9–11 mg/dL
C. 4–5 mEq/L
D. 96–106 mEq/L

39-10. Daily requirement of iron is:

A. 10–20 µg
B. 0.1–0.5 g
C. 0.1–0.5 mg
D. 10–20 mg

39-11. Iron is deficient in:

A. Cereals (rice, wheat)
B. Maize and corn
C. Pulses (bengal gram, etc.)
D. Milk

39-12. Which enzyme do not contain copper?

A. Cytochrome oxidase
B. Superoxide dismutase
C. Xanthine oxidase
D. Tyrosinase

39-13. Daily requirement of zinc for an adult is:

A. 2 mg
B. 10 mg
C. 40 mg
D. 70 mg

39-14. Zinc is present in all enzymes, *except:*

A. Alkaline phosphatase
B. Amylase
C. Carbonic anhydrase
D. Carboxy peptidase

39-15. Tetany can result from all, *except:*

A. Hypocalcemia
B. Vitamin D deficiency
C. Renal tubular acidosis
D. Hypomagnesemia

39-16. Which is not true regarding iron?

A. Body level is regulated by increasing excretion
B. Transported as ferric iron bound to transferrin
C. Absorbed as ferrous form only
D. Main storage is ferritin in liver

39-17. Daily requirement of iodine for an adult is:

A. 15–20 μg
B. 150–200 μg
C. 800–900 μg
D. 1–2 μg

39-18. The micromineral present in teeth is:

A. Calcium
B. Iodine
C. Fluorine
D. Manganese

39-19. Zinc is present in all the enzymes, *except:*

A. Alkaline phosphatase
B. Amylase
C. Carbonic anhydrase
D. RNA polymerase

39-20. Which of the following trace elements has an antioxidant role?

A. Chromium B. Zinc
C. Selenium D. Nickel

39-21. Accumulation of iron (hemosiderosis) leads to all the following clinical features, *except:*

A. Diabetes mellitus
B. Cirrhosis of liver
C. Defective coagulation
D. Hyper pigmentation

39-22. Which of the following is not a feature of Wilson's hepatolenticular degeneration?

A. Decreased excretion of copper in urine
B. Low ceruloplasmin level
C. Keyser-Fleischer ring in cornea
D. Cirrhosis of liver

39-23. Which of the following is not true regarding the absorption of calcium?

A. Absorption is increased by vitamin D
B. Acidity favors calcium absorption
C. Deficiency of bile favors absorption
D. Basic amino acids increase absorption

39-24. Parathyroid hormone regulates calcium homeostasis by all the mechanisms, *except:*

A. Bone resorption
B. Secretion of calcitonin
C. Increased absorption of calcium from intestine
D. Increased reabsorption from renal tubules

39-25. Hyperparathyroidism is characterized by all the following, *except:*

A. Osteoporosis B. Hypercalcuria
C. Alkalosis D. Recurrent urinary calculi

39-26. Sodium level in serum is affected by all the following factors, *except:*

A. Aldosterone B. Cortisone
C. ADH D. Cholecalciferol

39-27. Which contains iron?

A. Ceruloplasmin
B. Xanthine oxidase
C. Albumin
D. Superoxide dismutase

39-28. All are true with selenium, *except:*
A. Action of selenium is complementary to vitamin E
B. Glutathione peroxidase contains selenium
C. Selenium may cause hepatic necrosis
D. Requirement is 50–100 μg/day

39-29. In a patient the serum calcium level was 12 mg/dL, and serum inorganic phosphate was 2 mg/dL. The above findings are characteristic of
A. Excess calcitonin secretion
B. Vitamin D deficiency
C. Hyperparathyroidism
D. Hypervitaminosis D

39-30. Fasting samples give lower values than post-prandial samples for all the parameters, *except:*
A. Glucose B. Phosphate
C. Triglycerides D. Cholesterol

39-31. Hypoklemia occurs in all conditions, *except:*
A. Rapid correction of acidosis
B. Infusion of insulin and glucose
C. Use of loop diuretics
D. Polycythemia

Answers (Chapter 39)

39-1. B.	39-2. D.	39-3. B.	39-4. D.
39-5. C.	39-6. D.	39-7. D.	39-8. C.
39-9. D.	39-10. D.	39-11. D.	39-12. C.
39-13. B.	39-14. B.	39-15. A.	39-16. A.
39-17. B	39-18. C	39-19. B.	39-20. C.
39-21. D.	39-22. A.	39-23. C.	39-24. B.
39-25. C.	39-26. D.	39-27. B.	39-28. C
39-29. C	39-30. B	39-31. D.	

CHAPTER 40: NUTRITION

40-1. Negative nitrogen balance is observed in:
A. Pregnancy
B. Chronic fever
C. Convalescence
D. Growth period

40-2. Positive nitrogen balance observed in all the following conditions, *except:*
A. Pregnant woman of 25 years
B. Obese diabetic of 60 years
C. Young athlete of 15 years
D. 25-year-old man during convalescence after surgery

40-3. Lysine is deficient in:
A. Cereals (rice, wheat)
B. Maize and corn
C. Pulses (bengal gram, etc.)
D. Tapioca (cassava)

40-4. Methionine is deficient in:
A. Cereals (rice, wheat)
B. Maize and corn
C. Pulses (bengal gram, etc.)
D. Tapioca (cassava)

40-5. Which has the highest calorific value?
A. Glucose
B. Palmitic acid
C. Albumin
D. Ethanol

40.6. The respiratory quotient (RQ) of a person oxidizing mainly fat is:
A. 0.5
B. 0.7
C. 0.9
D. 1.0

40-7. Calories generated per gram of fat is:
A. 4 kcal
B. 5 kcal
C. 8 kcal
D. 9 kcal

40-8. Basal metabolic rate is increased by all the following, *except:*
A. Fever
B. Thyroxine
C. Starvation
D. Cold climate

40-9. The beneficial effects of dietary fiber include all the following, *except:*
A. Increased motility of intestine
B. Helping in the digestion process
C. Decreased absorption of cholesterol
D. Increased glucose tolerance

40-10. A balanced diet should have calories for carbohydrate, proteins and fats in the following ratio:
A. 50 : 15 : 35
B. 40 : 30 : 30
C. 60 : 20 : 20
D. 70 : 10 : 10

40-11. Calorie requirement of a 60 kg person doing sedentary work is approximately:

A. 1000 kcal B. 1500 kcal
C. 2000 kcal D. 2500 kcal

40-12. The nutritional quality of a protein is dependent on the following, *except:*

A. Content of essential amino acids
B. Digestibility
C. Relative proportion of amino acids
D. Energy yield from the protein

40-13. The daily nitrogen excretion in a normal healthy adult consuming 100 g of protein will be:

A. 8 g B. 10 g
C. 16 g D. 20 g

40-14. The minimum daily requirement of proteins in food for a normal healthy adult is:

A. 0.2 g/kg body weight
B. 0.4 g/kg
C. 0.7 g/kg
D. 0.9 g/kg

40-15. Phenyl alanine is deficient in:

A. Cereals (rice, wheat)
B. Maize and corn
C. Pulses (bengal gram, etc.)
D. Tapioca (cassava)

40-16. Which of is not an essential nutrient?

A. Sucrose B. Linoleic acid
C. Tryptophan D. Ascorbic acid

40-17. The following carbohydrate is undigestible:

A. Lactose B. Glycogen
C. Fructose D. Pectin

40-18. Which has the highest calorific value?

A. Glucose B. Palmitic acid
C. Albumin D. Ethanol

40-19. Which has no role in calculating calorie requirements?

A. Respiratory quotient
B. Specific dynamic action
C. Nature of work
D. Basal metabolic rate

40-20. Basal metabolic rate is increased by all the following, *except:*

A. Fever B. Thyroxine
C. Starvation D. Cold climate

40-21. Consumption of refined sugar can lead to all the following effects, *except:*

A. Sudden hyperglycemia
B. Hypertriglyceredemia
C. Glycogen accumulation
D. Dental caries

40-22. Which of the following pairs of food items will not provide all essential amino acids:

A. Fish and tapioca B. Chapathy and dal
C. Rice and curds D. Bread and jam

40-23. Which of the following is the best index of the nutritive value of a protein?

A. Biological value
B. Protein energy ratio
C. Net protein utilization
D. Net dietary protein value

40-24. Deficiency of proteins in the diet can lead to the following, *except:*

A. Positive nitrogen balance
B. Edema
C. Recurrent infection
D. Loss of appetite

40-25. Negative nitrogen balance may occur due to all the following conditions, *except:*

A. Vegetarian diet B. Cytotoxic therapy
C. Extensive burns D. Traumatic hemorrhage

40-26. All the following diseases may be associated with obesity, *except:*

A. Grave's disease
B. Cushing's syndrome
C. Depressive psychosis
D. Diabetes mellitus

40-27. All the following abnormalities are observed in obese people, *except:*

A. Decreased glucose tolerance
B. High insulin levels
C. Hypoalbuminemia
D. Hyperlipidemia

40-28. A child with Kwashiorkor may have all the following clinical features, *except:*
A. Hypoalbuminemia B. Hypercalcemia
C. Hypomagnesemia D. Normoglycemia

40-29. A person with 60 kg weight, was given food containing 65 g of proteins per day for two weeks. After the experiment, his weight was found to be 60 kg itself. His nitrogen balance is said to be:
A. Minus 5 g B. Minus 2.5 g
C. 0 g D. Plus 2.5 g

40-30. Nutritional status of a person can be assessed by all, *except:*
A. Serum protein estimation
B. Hemoglobin level
C. Gain or loss in weight
D. Mid arm circumference

Answers (Chapter 40)

40-1. B.	40-2. B.	40-3. A.	40-4. C.
40-5. B.	40-6. B.	40-7. D.	40-8. C.
40-9. B.	40-10. C.	40-11. C.	40-12. D.
40-13. C.	40-14. C.	40-15. D.	40-16. A.
40-17. D.	40-18. B	40-19. A.	40-20. C.
40-21. C.	40-22. D.	40-23. C.	40-24. A.
40-25. A.	40-26. A.	40-27. C.	40-28. B.
40-29. C.	40-30. D.		

CHAPTER 43: NUCLEOTIDES

43-1. All the following amino acids are used for synthesis of purine nucleotides, *except:*
A. Glycine B. Alanine
C. Aspartate D. Glutamine

43-2. N-3 of purine ring is donated by:
A. Ammonia B. Glutamine
C. Aspartate D. Glycine

43-3. Which amino acid is required for both purine and pyrimidine synthesis?
A. Glycine B. Aspartate
C. Alanine D. Glutamate

43-4. The chief product of catabolism of purines in human beings is:

A. Urea
B. Uric acid
C. Hypoxanthine
D. Beta aminoisobutyric

43-5. Hyperuricemia is observed in all the following conditions, *except:*

A. Lesch-Nyhan syndrome
B. Von Gierke's Disease
C. Glutathione reductase deficiency
D. Xanthine oxidase deficiency

43-6. Hyperuricemia can result from deficiency of all the following enzymes, *except:*

A. HGPRTase
B. APRTase
C. Glucose-6-phosphatase
D. Adenosine deaminase

43-7. The reduction of nucleotide to deoxyribonucleotide:

A. Takes place at the level of nucleoside diphosphate
B. Involves the cleavage of glycosidic bond
C. d-ATP will enhance the reaction
D. Needs hydrolysis of ATP

43-8. Orotic aciduria is a feature of deficiency of all the following enzymes, *except:*

A. OMP decarboxylase
B. Ornithine transcarbamoylase
C. OPRTase
D. Aspartate transcarbamoylase

43-9. Pyrimidine nucleotide synthesis and purine nucleotide synthesis are similar in which aspects?

A. Enzymes are present as multienzyme complexes
B. Ring is assembled and ribose-5-phosphate added
C. Ribose-5-phosphate is donated by PRPP
D. Aspartic acid is fully incorporated into the ring

43-10. All the following molecules have regulatory effect on synthesis of purine nucleotides, *except:*

A. PRPP B. AMP
C. UMP D. GTP

43-11. Which property of uric acid is responsible for the manifestations of gout?

A. Ketoenol tautomerism
B. Acidic nature
C. Reducing action
D. Solubility constant

43-12. Which compound is not derived from purine nucleotides?

A. Uric acid
B. Beta amino isobutyric acid
C. Allantoin
D. Xanthylic acid

43-13. Uric acid is the end product of breakdown of all the following, *except*:

A. AMP B. IMP
C. Xanthosine D. CMP

43-14. Secondary hyperuricemia is seen in all the following conditions, *except*:

A. Leukemia B. Psoriasis
C. Hepatitis D. Cytotoxic drug therapy

43-15. Which is correct regarding thymine synthesis?

A. 4th carbon is donated by carbamoyl phosphate
B. 3rd nitrogen is donated by aspartate
C. Methylation is taking place on dUMP and not on UMP
D. 1st nitrogen is donated by glutamine

43-16. Regarding gout, all are true, *except:*

A. It is seen in HGPRT deficiency
B. There is deposition of urate in joints
C. Can be treated using allopurinol
D. It is seen in PRPP synthetase deficiency

43-17. If a cell is unable to synthesize PRPP, synthesis of which is likely to be directly affected?

A. Pyridoxal B. OMP
C. CoASH D. FAD

43-18. During de novo synthesis of pyrimidine, which nucleotide is first formed?

A. TMP B. OMP
C. UMP D. CMP

43-19. The drug of choice for primary gout is:

A. Allopurinol B. Aspirin
C. Colchicine D. Probenecid

43-20. All manifestations are seen in Lesch Nyhan syndrome, *except:*

A. Self-mutilation B. Imunodeficiency
C. Hyperuricemia D. X-linked inheritance

43-21. All are true regarding pyrimidine nucleotide synthesis, *except:*

A. The reactions occur in the cytoplasm
B. CTP is a negative modifier
C. Glutamine and aspartic acid are used
D. Pyrimidine ring is built up on a ribose-5-phosphate

43-22. Formation of dTMP requires:

A. S-adenosyl methionine
B. NADPH
C. Adrenodoxin
D. Molybdenum

43-23. Orotic aciduria is characterized by all the findings, *except:*

A. Megaloblastic anemia
B. Response to oral uridine
C. Failure to thrive
D. Deficiency of HGPRTase

43-24. Which of the following conditions is not associated with hyperuricemia?

A. Multiple myeloma B. Xanthinuria
C. Psoriasis D. Polycythemia vera

Answers (Chapter 43)

43-1. B.	43-2. B.	43-3. B.	43-4. B.
43-5. D.	43-6. D.	43-7. A.	43-8. D.
43-9. C.	43-10. C.	43-11. D.	43-12. B.
43-13. D.	43-14. C.	43-15. C.	43-16. A.
43-17. B.	43-18. B.	43-19. A.	43-20. B.
43-21. D.	43-22. B.	43-23. D.	43-24. B.

CHAPTER 44: DNA STRUCTURE AND REPLICATION

44-1. Which nucleoside is found in DNA?
A. Pseudouridine B. Dihydrouridine
C. Deoxy thymidine D. Inosine

44-2. Nucleosome is:
A. Synonym of nucleolus
B. DNA-RNA complex present in nucleus
C. mRNA attached with snRNA
D. DNA wrapped around histones

44-3. Fidelity of replication is ensured by:
A. Complementary base pairing
B. Specificity of DNA polymerase
C. Recognition of specific nucleotide sequence
D. Formation of phospho diester bonds

44-4. During DNA replication, 5'-GTCAAT-3' strand produces a new complementary DNA strand with the following structure:
A. 5'-CAGTTA-3' B. 5'-ATTGAC-3'
C. 5'-ACACCG-3' D. 5'-GCCACA-3'

44-5. All are due to defective DNA repair, *except:*
A. Xeroderma pigmentosum
B. Fanconi's anemia
C. Progeria
D. Cystic fibrosis

44-6. Which enzyme protects DNA from ageing?
A. DNA polymerase
B. Topoisomerase
C. Deoxyribonuclease
D. Telomerase

44-7. Which is not true in xeroderma pigmentosum?
A. It is an autosomal recessive condition
B. Defect is in the repair mechanism of DNA, especially thymine dimers
C. Sensitivity to UV light is the major manifestation
D. Cells are unable to synthesize vitamin D even in presence of sunlight

44-8. In prokaryotes, DNA replication is inhibited by:
A. Rifampicin B. Ciprofloxacine
C. Erythromycin D. Streptomycin

44-9. DNA synthesis in eukaryotes is affected by the following, *except:*
A. Cytosine arabinoside
B. 5-fluorouracil
C. Rifampicin
D. 6-mercaptopurine

44-10. All the following statements are true with regard to Watson-Crick model of DNA, *except:*
A. Double stranded helix
B. DNA strands running in opposite directions
C. Adenine bonds with thymine
D. Guanine bonds with uracil

44-11. All are true with regard to histones, *except:*
A. They are basic proteins
B. Nucleosomes contain histones and DNA
C. Phosphorylation increases transcriptional activity
D. Histone synthesis is coupled with DNA synthesis

44-12. Histones are modified by all, *except:*
A. ADP ribosylation B. Acetylation
C. Glycosylation D. Phosphorylation

44-13. Transcriptionally active chromatin is:
A. Resistant to digestion by deoxyribonuclease
B. Methylated at cytosine residues
C. Densely packed
D. Called euchromatin

44-14. DNA replication requires all, *except:*
A. Elongation factor
B. Deoxynucleotide phosphates
C. DNA polymerase
D. RNA primer

44-15. Which enzyme is not required for replication?
A. DNA ligase B. Topoisomerase
C. Helicase D. Reverse transcriptase

44-16. All are involved in synthesis of DNA, *except:*
A. DNA polymerase
B. Topoisomerase
C. Reverse transcriptase
D. Deoxyribonuclease

44-17. All cause fidelity in replication, *except:*
A. There is an inherent repair mechanism
B. DNA polymerase beta has an editing function
C. DNA polymerase also has exonuclease activity
D. Replication stops if a wrong base is added

44-18. Regarding DNA structure which is not correct?
A. Two strands are covalently bonded
B. Two strands are antiparallel
C. Strands have polarity
D. Strands are complementary to each other

44-19. Which of the following is not a feature of mitochondrial myopathies?
A. Cytoplasmic inheritance
B. Heteroplasmy
C. Genomic imprinting
D. Low energy threshold

44-20. All are regulatory sequences on DNA, *except:*
A. Attenuator sequences
B. Enhancer sequences
C. Consensus sequences
D. Promoter sequences

44-21. Denaturation of DNA is accompanied by all the changes, *except:*
A. Increase in viscosity
B. Hyperchromicity
C. Unstacking of bases
D. Destabilization of DNA helix

Answers (Chapter 44)

44-1. C.	44-2. D.	44-3. A.	44-4. B.
44-5. D.	44-6. D.	44-7. D.	44-8. B.
44-9. C.	44-10. D.	44-11. C.	44-12. C.
44-13. D.	44-14. A.	44-15. D.	44-16. D.
44-17. D.	44-18. A.	44-19. A.	44-20. C.
44-21. A.			

CHAPTER 45: TRANSCRIPTION

45-1. All the bases are found in mRNA, *except:*
A. Adenine B. Guanine
C. Uracil D. Thymine

45-2. The nitrogenous base absent in DNA is:

A. Uracil
B. Cytosine
C. Thymine
D. Guanine

45-3. Sigma factor is:

A. A subunit of DNA polymerase
B. A subunit of RNA polymerase
C. A subunit of 50S ribosome
D. Responsible for initiation of replication

45-4. The processing of mRNA involves

A. Capping at 5' terminus
B. Poly A tail attachment
C. Removal of introns
D. Translation of mRNA into nucleus

45-5. Intron is the portion of:

A. DNA that is cleaved off during replication
B. mRNA that is removed after transcription
C. tRNA that is added on after its synthesis
D. Protein that is removed after translation

45-6. Which is an inhibitor of transcription in eukaryotes?

A. Rifampicin
B. Chloramphenicol
C. Alpha amanitin
D. Streptomycin

45-7. Mechanism of action of mitomycin is:

A. Inactivation of EF-2; blocks elongation of proteins
B. Intercalation with DNA strands; blocks transcription
C. Abrupt termination of protein synthesis
D. It is a structural analog of amino acyl tRNA

45-8. Reverse transcriptase differs from DNAP in that:

A. Synthesises DNA in 5´ to 3´ direction
B. Has 3' to 5 ' exonuclease activity
C. Follows the base paairing rule
D. Can insert iIMP in the growing strand

45-9. RNAP is inhibited by:

A. Streptomycin
B. Chloramphenicol
C. Ciprofloxacin
D. Rifampicin

45-10. Which of the following nucleotides is not present in mRNA?

A. GMP
B. TMP
C. CMP
D. UMP

45-11. Which of the enzymes listed below is involved in processing of RNA?

A. RNA polymerase II B. Ligase
C. Helicase D. Ribonuclease P

45-12. All the following are inhibitors of eukaryotic transcription, *except*:

A. Rifampicin B. Puromycin
C. Actinomycin D. Amanitin

45-13. The coding unit of the DNA is called as:

A. Intron B. Exon
C. Cistron D. Prion

45-14. Which is not a transcription start signal?

A. CAAT box B. Hogness box
C. Pribnow box D. Homeo box

45-15. All are true regarding transcription, *except*:

A. RNA polymerase has proof reading capacity
B. RNA strand is complementary to template strand of DNA
C. Is inhibited by rifampicin
D. RNA is synthesized in the 5´ to 3´ direction

45-16. In an organism, only DNA, but not RNA is labelled when grown in a medium containing radioactive nucleotide of:

A. Adenine B. Guanine
C. Thymine D. Cytosine

45-17. The RNA present in largest amounts in a cell at any time is:

A. mRNA B. tRNA
C. rRNA D. hnRNA

45-18. All the bases are found in mRNA, *except*:

A. Adenine B. Guanine
C. Uracil D. Thymine

45-19. The nitrogenous base absent in DNA is:

A. Uracil B. Cytosine
C. Thymine D. Guanine

45-20. The TATA box:

A. Present on the template strand
B. Present about 70 bp away from transcription start site
C. Serves as the signal for attachment of RNAP-II
D. Acts as a silencer of the gene

45-21. The processing of mRNA involves:
A. Capping at 5′ terminus
B. Poly A tail attachment
C. Removal of introns
D. Translation of mRNA into nucleus

45-22. All are post-transcriptional changes, *except*:
A. Modification of bases
B. Splicing of introns
C. Trimming of 3′ end to CCA sequence
D. Attachment of poly A tail

45-23. Transcription is terminated when:
A. The codon on the mRNA is AUG
B. Rho factor binds to mRNA
C. The codon on the mRNA is UGA
D. The ribosome reaches the poly-A tail on mRNA

45-24. Which is caused by autoantibodies to Snurps?
A. Xeroderma pigmentosum
B. Mushroom toxicity
C. Systemic lupus erythematosus
D. Thalassemia

45-25. Intron is the portion of:
A. DNA that is cleaved off during replication
B. mRNA that is removed after transcription
C. tRNA that is added on after its synthesis
D. Protein that is removed after translation

45-26. All are correct with retroviruses, *except*:
A. Contains a DNA dependent RNA polymerase
B. Transcription occurs in the reverse direction
C. Viral RNA is copied into host cell DNA
D. New DNA is incorporated into host cell genome.

45-27. Mechanism of action of mitomycin is:
A. Inactivation of EF-2, and so blocks elongation
B. Intercalation of DNA strands and blocks transcription
C. Abrupt termination of protein synthesis
D. It is a structural analogue of amino acyl tRNA

Answers (Chapter 45)

45-1. D.	45-2. A.	45-3. B.	45-4. B.
45-5. B.	45-6. C.	45-7. B.	45-8. D.
45-9. B.	45-10. B.	45-11. D.	45-12. A.
45-13. C.	45-14. D.	45-15. A.	45-16. C.
45-17. C.	45-18. D.	45-19. A.	45-20. C.
45-21. B.	45-22. C.	45-23. B.	45-24. C.
45-25. B.	45-26. A.	45-27. B.	

CHAPTER 46: TRANSLATION

46-1. Degeneracy of genetic code denotes:

A. Single codon stands for multiple amino acids
B. One amino acid is represented by multiple codons
C. Overlapping of codons is observed at times
D. Some codons contain 4 bases instead of 3 bases

46-2. The term wobbling means:

A. Occasional error in translation
B. Low fidelity in mRNA synthesis
C. The ability of anticodons to pair with codons that differ at the third base
D. Post-transcriptional modification of excision of introns

46-3. Which is not true regarding the genetic code:

A. Degenerate B. Ambiguous
C. Non-overlapping D. Universal

46-4. During protein biosynthesis, high energy bonds are utilized in all the following steps, *except*:

A. Formation of amino acyl tRNA
B. Binding of amino acyl tRNA to A site of mRNA ribosome complex (EF-1 step)
C. Formation of peptide bond (peptidyl transferase step)
D. Translocation (EF-2 step)

46-5. Post-translational modifications include all the following, *except*:

A. Glycosylation B. Hydroxylation
C. Decarboxylation D. Phosphorylation

46-6. All are diseases resulting from failure of post-translational modifications, *except:*
A. Lathyrism
B. Ehler-Danlos syndrome
C. Osteogenesis imperfecta
D. Scurvy

46-7. Failure of post-translational modification occurs in deficiency of which vitamin?
A. Ascorbic acid
B. Retinol
C. Vitamin B_{12}
D. Vitamin D

46-8. The mechanism of action of tetracycline is:
A. Inhibits tRNA binding to ribosome
B. Decreases binding of ribosomes to mRNA
C. Causes misreading of codes
D. Inhibits translocation

46-9. All are true regarding tRNA, *except:*
A. Amino acid is attached to 3´ end
B. Anticodon sequence is at 5´ end
C. Contains several modified bases
D. Has a clover leaf pattern due to hydrogen bonding

46-10. Unusual nucleotide bases are found in significant quantities in:
A. mRNA B. tRNA
C. rRNA D. snRNA

46-11. All are inhibitors of RNA synthesis, *except:*
A. Rifampicin B. Puromycin
C. Amanitin D. Actinomycin D

46-12. All are structural features of tRNA, *except:*
A. Has clover leaf appearance
B. Has amino acid binding site at 5´ end
C. Contains unusual bases
D. Has a pseudouridine arm

46-13. All the following are inhibitors of translation in prokaryotic cells, *except:*
A. Clindamycin
B. Streptomycin
C. Mitomycin
D. Erythromycin

46-14. Which of the following post-translational events is defective in Mad cow disease?
A. Amino terminal modification
B. Protein folding
C. Phosphorylation of serine residues
D. Glycosylation of asparagine

46-15. In eukaryotes the initiation of translation requires all the following, *except*:
A. Methionyl tRNA
B. 70 S ribosome
C. Shine-Dalgarno sequence
D. mRNA codon AUG

46-16. Elongation of polypeptide chain requires all the following proteins, *except*:
A. Peptidyl transferase
B. Rho factor
C. Translocase
D. GTPase

46-17. Which complex is involved in translation?
A. Polysome B. Nucleosome
C. Spliceosome D. Lysosome

46-18. Signal peptide (SP) region is necessary for all the following, *except*:
A. Attachment of ribosomes to mRNA
B. Anchorage of ribosomes to endoplasmic reticulum
C. Release of newly synthesized proteins from ribosomes
D. Digestion of signal recognition particle (SRP)

46-19. Which is not a post-translational modification?
A. Gamma carboxylation of prothrombin
B. Hydroxylation of proline in collagen
C. Methylation of histones
D. Hydroxylation of benzene ring of phenylalanine

46-20. Which is an inhibitor of translation in eukaryotic cells?
A. Erythromycin
B. Puromycin
C. Chloramphenicol
D. Tetracycline

46-21. Mechanism of action of streptomycin is that it:
A. Inhibits tRNA binding to ribosome
B. Decreases binding of ribosomes to mRNA
C. Causes misreading of codes
D. Inhibits translocation

46-22. All the following statements are true with regard to mitochondrial DNA, *except*:
A. All mitochondrial proteins are encoded by mitochondrial DNA
B. Mitochondrial DNA is inherited by cytoplasmic inheritance.
C. Inheritance is predominantly from mother
D. DNA in mitochondria is circular

46-23. The disease resulting from mutation in mitochondrial DNA is:
A. Zellweger syndrome
B. Jakob Creutzfeldt disease
C. Xeroderma pigmentosum
D. Leber's optic myopathy

46-24. Which is due to defective protein targetting?
A. Xeroderma pigmentosum
B. Thalassemia
C. Primary oxaluria
D. Ataxia telengectasia

Answers (Chapter 46)

46-1. B.	46-2. C.	46-3. B.	46-4. C.
46-5. C.	46-6. C.	46-7. A.	46-8. A.
46-9. B.	46-10. B.	46-11. B.	46-12. B.
46-13. C.	46-14. B.	46-15. B.	46-16. B.
46-17. A.	46-18. D.	46-19. D.	46-20. B.
46-21. C.	46-22. A.	46-23. D.	46-24. B.

CHAPTER 47: GENE EXPRESSION, MUTATION

47-1. All the following traits are inherited through the X-chromosome, *except*:
A. Sickle cell anemia
B. Hemophilia
C. Glucose-6-phosphate dehydrogenase deficiency
D. Duchenne type muscular dystrophy

47-2. In autosomal dominant inheritance, if only one parent is affected:

A. 50% of children will be sufferers
B. 25% of children will be carriers
C. 25% of children will be normal
D. All the children will be genotypically abnormal

47-3. All the following lead to cancer, *except*:

A. Promoter insertion
B. Chromosomal translocation
C. Point mutation
D. Trinucleotide expansion

47-4. Point mutations may be silent because:

A. A change in third base of the codon may produce a synonym
B. Change in base sequence is corrected during transcription
C. Non-specific base pairing between mRNA and DNA
D. High fidelity of transcription of mRNA

47-5. HbS (sickle cell anemia) is a result of:

A. Transversion mutation
B. Transition mutation
C. Nonsense mutation
D. Frameshift mutation

47-6. Mutagenicity of a compound is checked by:

A. Hay's test B. Guthrie test
C. Ames' test D. FISH test

47-7. In which phase of the cell cycle synthesis of deoxyribonucleotide triphosphate is highest?

A. G0 B. G1
C. S D. M

47-8. Which hormone does not act at the level of transcription?

A. Cortisol B. Calcitriol
C. Aldosterone D. Calcitonin

47-9. Viruses may have all the following fates within the host cells, *except*:

A. Virus particles may multiply within host cells
B. Viral DNA may be incorporated into host cell DNA
C. Viral DNA may be transcribed
D. Viruses may suppress oncogenes

47-10. All the following are antiviral drugs, *except*:
A. Cytosine arabinoside
B. Acyclovir
C. Ribavarin
D. Azidothymidine

47-11. In the process of transduction:
A. Bacteria takes up DNA from the surrounding medium
B. New genes are produced by accumulation of mutations
C. Normal cell is transformed into a malignant cell
D. Virus carries some part of the genetic information from one host to another host

47-12. All are correct with retroviruses, *except*:
A. Can use their RNA as template
B. Have reverse transcriptase
C. Can multiply in host cells
D. Are inactivated after the first multiplication cycle

47-13. Frameshift mutation results from:
A. Substitution of a single base
B. Deletion of a single base
C. Addition of a codon
D. Deletion of a codon

47-14. Abnormal genotype and normal phenotype is seen in mutation of:
A. Splice site B. Intervening sequences
C. Terminator codon D. Transcription factors

47-15. When GAC is replaced by GAA, the result is:
A. Back mutation B. Deletion mutation
C. Silent mutation D. Conservative mutation

47-16. The cell cycle is regulated by:
A. Availability of nucleotides
B. Presence of inducers in cells
C. Cyclin dependent kinase
D. DNA dependent kinases

Answers (Chapter 47)

47-1. A.	47-2. A.	47-3. D	47-4. A.
47-5. A.	47-6. C.	47-7. C.	47-8. D.
47-9. D.	47-10. A	47-11. D.	47-12. D.
47-13. B.	47-14. B.	47-15. C.	47-16. C.

CHAPTER 48: RECOMBINANT DNA

48-1. Which enzyme is used for preparing a recombinant DNA molecule?
A. Restriction endonuclease
B. RNA polymerase
C. DNA polymerase
D. Topoisomerase

48-2. Application of recombinant DNA technology include all the following, *except:*
A. Detection of oncogenes
B. Synthesis of human proteins
C. Inhibition of replication
D. Gene therapy

48-3. With plasmids, all are correct, *except:*
A. Used as vectors in recombinant DNA technology
B. Carriers of genes that confer drug resistance
C. Extra chromosomal duplex DNA molecules
D. Specialized regions of bacterial chromosomes

48-4. With plasmids, all are correct, *except:*
A. They are useful for ELISA technique
B. They are circular double stranded DNA inside bacteria
C. They replicate independent of bacterial DNA replication
D. They confer antibiotic resistance to host bacteria

48-5. All the following are nucleic acids, *except:*
A. Plasmids B. Prions
C. Cosmids D. Virions

48-6. Synthesis of recombinant DNA (rDNA) requires all the following, *except:*
A. Restriction endonuclease
B. RNA primer
C. Plasmid vector
D. DNA ligase

48-7. Which is not amenable to gene therapy?
A. Hemophilia
B. Severe combined immunodeficiency
C. Cystic fibrosis
D. Infective hepatitis

48-8. All are vectors for gene therapy, *except*:
A. Proteasomes B. Liposomes
C. Adenoviruses D. Retroviruses

48-9. The cDNA is prepared by using the enzyme:
A. RNA polymerase
B. DNA polymerase
C. Reverse transcriptase
D. Restriction endonuclease

48-10. Which of the following is not used as a vector in molecular cloning?
A. Plasmid
B. Cosmid
C. Artificial chromosome
D. Hapten

48-11. Restriction endonucleases:
A. Cleaves DNA at splice sites
B. Protects bacterial genome
C. Hydrolyzes DNA from 5′ end
D. Inhibits human DNAP

48-12. The enzyme deficiency which was first corrected by gene therapy is:
A. Adenosine deaminase
B. HGPRTase
C. Glucose-6-phosphatase
D. APRTase

Answers (Chapter 48)

48-1. A.	48-2. C.	48-3. D.	48-4. A.
48-5. B.	48-6. B.	48-7. D.	48-8. A.
48-9. C.	48-10. D.	48-11. B.	48-12. A.

CHAPTER 49 : MOLECULAR BIOLOGY TECHNIQUES

49-1. Specific DNA can be identified by all the following techniques, *except*:
A. Southern blotting B. In situ hybridization
C. Western blotting D. Northern blotting

49-2. Southern blot technique involves all, *except*:
A. Cleavage of DNA by restriction endonuclease
B. Addition of radiolabelled antibody for detection
C. Electrophoresis of fragments
D. Blot on nitro-cellulose sheet

49-3. A radiolabelled DNA probe is usually employed for all the following processes, *except:*
A. Southern blotting B. Northern blotting
C. DNA hybridization D. Western blotting

49-4. RFLP (restriction fragment length polymorphism) is used to:
A. Identify a specific gene in bacteria
B. Locate mutations in DNA
C. Study the rate of transcription
D. To amplify genes

49-5. Which exhibits RFLP ?
A. Haptoglobin B. Insulin
C. Beta globin chain D. Myoglobin

49-6. Which technique has medicolegal application?
A. DNA hybridization B. DNA fingerprinting
C. DNA denaturation D. DNA transposition

49-7. All the following are requisites for PCR, *except:*
A. DNA primer
B. Taq polymerase
C. Restriction endonuclease
D. Deoxynucleotide triphosphates

49-8. Gene amplification can be achieved by:
A. Southern blotting
B. DNA finger printing
C. Polymerase chain reaction
D. Gene cloning

49-9. Which of the following techniques is not useful in prenatal diagnosis?
A. DNA probing
B. Linkage analysis
C. DNA finger printing
D. DNA looping

49-10. Which is involved in renaturation of DNA?
A. Excision B. Splicing
C. Annealing D. Priming

49-11. PCR is used for the following, *except:*
A. Diagnosis of tuberculosis
B. Establishment of biological parentage
C. Detection of oncogenes
D. Detection of anti-nuclear antibodies

49-12. Which cannot detect specific DNA sequences?
A. FISH
B. ELISA
C. Dot blot hybridization
D. Autoradiography

Answers (Chapter 49)

49-1. C.	49-2. B.	49-3. D.	49-4. B.
49-5. A.	49-6. B.	49-7. C.	49-8. .C.
49-9. D.	49-10. C.	49-11. D.	49-12. B.

CHAPTER 50: HORMONES, MECHANISM OF ACTION

50-1. All the following compounds listed below may act as second messengers, *except:*
A. Cyclic AMP B. 2,3-bisphosphoglycerol
C. Calcium ions D. Inositol triphosphate

50-2. All the following hormones have membrane receptors, *except:*
A. Insulin B. Epinephrine
C. Glucagon D. Thyroxine

50-3. All the hormones listed below act at the level of transcription, *except:*
A. Cortisol B. Calcitriol
C. Calcitonin D. Androgen

Answers (Chapter 50)

50-1. B. 50-2. D. 50-3. C.

CHAPTER 51: HYPOTHALAMIC HORMONES

51-1. One of the following is not a glycoprotein:
A. HCG B. TSH
C. GnRH D. FSH

51-2. Alpha chain of human chorionic gonadotropin (hCG) is shared with all the following, *except*:
A. TSH B. FSH
C. LH D. ACTH

51-3. Growth hormone secretion is decreased:
A. Following exercise B. During sleep
C. Hyperglycemia D. Arginine infusion

Answers (Chapter 51)

51-1. B. 51-2. D. 51-3. C.

CHAPTER 52: STEROID HORMONES

52-1. All are true with regard to aldosterone, *except:*
A. Has 21 carbon atoms
B. It is the most active mineralocorticoid
C. Increases renal tubular reabsorption of sodium
D. Produced by adrenal medulla

52-2. 11-hydroxylase is required for the synthesis of:
A. Cortisol B. Estradiol
C. Testosterone D. Epinephrine

52-3. 17-hydroxylase is required for the synthesis of all the following, *except:*
A. Cortisol B. Estradiol
C. Testosterone D. Aldosterone

52-4. 21-hydroxylase is required for the synthesis of:
A. Estradiol B. Testosterone
C. Aldosterone D. Epinephrine

52-5. All are derived from progesterone, *except*:
A. Corticosterone
B. Aldosterone
C. Testosterone
D. Dehydroepiandrosterone

52-6. All are effects of glucocorticoids, *except:*
A. Stimulates immune function
B. Increased gluconeogenesis
C. Augmented lipolysis
D. Elevated protein breakdown

52-7. Steroid hormones are produced from:
A. Purine
B. Pyrimidine
C. Cholesterol
D. Stercobilinogen

52-8. The assessment of fetoplacental unit is done by measuring urinary:
A. Pregnanetriol
B. 17-hydroxysteroids
C. Estriol
D. Estradiol

52-9. Which of the tests listed below is used as a screening test for Cushing's syndrome?
A. Midnight cortisol level in plasma
B. Urinary free cortisol
C. Overnight dexamethasone suppression test
D. Plasma ACTH level

Answers (Chapter 52)

52-1. D.	52-2. A.	52-3. D.	52-4. C.
52-5. A.	52-6. A.	52-7. C.	52-8. C.
52-9. C.			

CHAPTER 53: THYROID HORMONES

53-1. Which is a feature of hypothyroidism?
A. Decreased T3 levels in blood
B. Decreased TSH levels
C. Weight loss
D. Increased basal metabolic rate

53-2. Primary hyperthyroidism is characterized by:
A. Low TSH, high T3 and high T4 levels
B. High TSH, T3 and T4 levels
C. Low TSH, T3 and T4 levels
D. High TSH, low T3 and low T4 levels

53-3. Thiouracil decreases thyroid function by:
A. Decreasing iodine uptake by thyroid gland
B. Reducing oxidation of iodine to active iodide
C. Diminishing iodination of thyroglobulin
D. Lowering coupling of di-iodo tyrosine to T4

53-4. Uptake of iodine by thyroid gland inhibited by:
A. Fluoride B. Perchlorate
C. Bromide D. Chloride

53-5. TSH stimulates all functions, *except:*
A. Uptake of iodine by thyroid gland
B. Oxidation of iodine to iodide by thyroid cell
C. Hydrolysis of thyroglobulin to produce T4
D. Deiodination of T4 to T3

53-6. Deficiency of thyroxine results in:
A. Grave's disease
B. Myxedema
C. Cushing's syndrome
D. Thyrotoxicosis

53-7. Investigations in suspected primary hypothyroidism include the following, *except:*
A. T4 B. TSH
C. TSIg D. Antithyroid antibodies

Answers (Chapter 53)

53-1. A. 53-2. A. 53-3. B. 53-4. B.
53-5. D. 53-6. B. 53-7. C.

CHAPTER 57: CANCER

57-1. All the following viruses are reported to have oncogenic, *except:*
A. Human papilloma virus
B. Varicella
C. Epstein-Barr virus
D. Hepatitis B virus

57-2. Oncogenes are specific sequences in DNA:
A. Which are tumor markers
B. That code for reverse transcriptase
C. When expressed produce cancer
D. Trigger off differentiation process

57-3. All are correct with src oncogene, *except:*
A. src DNA is seen in tumor tissue; not in normal cells
B. src gene is part of Rous sarcoma RNA virus
C. Oncogene product is localized in membrane
D. src product is a protein kinase that phosphorylates tyrosine

57-4. All are oncogene products, *except:*
A. Growth factors
B. Interleukin 2
C. Tumor necrosis factor
D. Tyrosine kinase

57-5. Oncogenes may be activated by all, *except:*
A. Viral infection
B. Promoter insertion
C. Mutations in proto-oncogene
D. Reverse transcriptase

57-6. Which malignancy results from the deletion of onco-suppressor gene?
A. Lymphosarcoma
B. Retinoblastoma
C. Multiple myeloma
D. Bronchogenic carcinoma

57-7. Which is true with regard to p53?
A. It is an oncogene
B. It is an oncosuppressor gene product
C. It is an inducer of heme synthesis
D. It is deleted in mental retardation

57-8. Alpha-fetoprotein level in serum is increased:
A. Prostatic cancer B. Hepatoma
C. Cancer lungs D. Nephritis

57-9. All the following are tumor markers, *except*:
A. Alpha-fetoprotein
B. Tartrate labile acid phosphatase
C. Regan isozyme of alkaline phosphatase
D. Alpha 1 antitrypsin

57-10. Competitive inhibitors of all the enzymes may be used as therapeutic agents, *except*:
A. Thymidylate synthase
B. Dihydrofolate reductase
C. Xanthine oxidase
D. Ribonucleotide reductase

57-11. Which is an anticancer drug?
A. Zidovudine B. 6-mercaptopurine
C. Allopurinol D. Acyclovir

57-12. Which of the diseases listed below is due to an onco-suppressor mutation?
A. Cystic fibrosis
B. Huntington's chorea
C. Sickle cell disease
D. Retinoblastoma

Answers (Chapter 57)

57-1. B.	57-2. C.	57-3. A.	57-4. B.
57-5. D.	57-6. B.	57-7. B.	57-8. B.
57-9. D.	57-10. D.	57-11. B.	57-12. D.

Index

Page numbers followed by *t* refer to table

NOTES

NOTES

NOTES

NOTES

NOTES

NOTES

NOTES

NOTES

NOTES

NOTES

NOTES

NOTES

NOTES

NOTES